The Challenge of Computing Homotopy Groups: Why Spheres Remain a Mystery

Sheena

Table of Contents

Chapter 1

Introduction

Algebraic topology has its roots in constructing, computing and understanding the deeper structure of three main invariants classically associated to a topological space X: homotopy groups $\pi_* X$, homology groups $H_* X$, and cohomology groups $H^* X$. While not always an easy task, computing homology and cohomology groups can often be done in purely algebraic terms given a nice enough presentation of your space X. Homotopy groups on the other hand are more elusive. Elements of the n-th homotopy group $\pi_n X$ are represented by maps from the n-dimensional sphere S^n into your space of choice X, up to continuous deformation (i.e., *homotopy*). In this elusiveness lies a more finessed invariant of spaces, as the homotopy groups often hold deeper information which homology and cohomology can not distinguish. The tradeoff then is that homotopy groups are generally much more difficult to compute, even providing a full description of the homotopy groups of the 2-sphere S^2 is an exceptionally difficult problem and still not fully understood.

Stable homotopy groups

As a result, it can be useful to know even an approximation of the homotopy groups of a space. The *stable homotopy groups* $\pi_*^s X$ of a space X are one such approximation. It follows from the Freudenthal suspension theorem [47, Corrolary 4.24] that if X is a k-connected space[1] that $\pi_n X \cong \pi_n^s X$ for $n \leq 2k$; i.e., that the stable homotopy groups agree with the homotopy groups in a "stable range". Stable homotopy groups are an instance of the homotopy groups of a more general type of algebro-topological object called a *spectrum* (see, e.g. [36], [51]) which is akin to a chain complex of spaces. Many computations involving spectra can be carried out by more formal algebra than their space-level compatriots, and as such the stable homotopy groups of a space are often significantly computable than ordinary homotopy groups.

But how are the stable homotopy groups constructed? The reader familiar with homology theory will recall that there is always an isomorphism

$$H^* X \cong H^{*+1}(\Sigma X) \tag{1.0.1}$$

where ΣX denotes the *suspension* of a space X (which may be modelled as the smash product in spaces of the 1-sphere and X, $\Sigma X = S^1 \wedge X$). Freudenthal's suspension theorem tell us that, unlike homology groups, the homotopy groups of a space only obey this isomorphism inside the stable range, i.e., $\pi_n X \cong \pi_{n+1}(\Sigma X)$ for $n \leq 2k$ when X is k-connected. However, in spaces the smash product $\wedge$ admits a right adjoint $Y \mapsto \Omega Y = \mathrm{Map}(S^1, Y)$[2]. Using the identification $\pi_{n+1}(\Sigma X) \cong [S^{n+1}, S^1 \wedge X]$, the isomorphism above is obtained by applying the n-th homotopy group functor to the *point-set level* map $\eta_X : X \to \Omega\Sigma X$. Iterating this map builds the *stabilization* QX of a topological space X as the following colimit whose maps are given by $\Omega^n \eta_{\Sigma^n X}$

$$QX := \Omega^\infty \Sigma^\infty X = \mathrm{colim}(X \to \Omega\Sigma X \to \Omega^2\Sigma^2 X \to \Omega^3\Sigma^3 X \to \cdots).$$

---

[1] This means that the groups $\pi_n X$ for $n \leq k$ are trivial

[2] The latter here is topologized by the compact-open topology, though ΩY can more easily be thought of as the *space* of loops in Y based at $* \in Y$.

The stable homotopy groups are obtained then as the ordinary homotopy groups of QX, i.e.,

$$\pi_*^s X := \pi_*(QX).$$

What's more is that, in the language of Goodwillie's calculus of homotopy functors, the functor Q may be thought of as a *first order*—or *linear*—approximation to the identity functor on based topological spaces. This strikes the question: What does linearity mean in this context?

Linear functors

Inspiration may be taken again from homology. Again, the familiar reader may recall that homology groups satisfy *excision* [47, Theorem 2.20] which gives rise to a strong computational tools such as the Mayer-Vietoris sequence [47, §2.22]. An equivalent statement is that any pushout square

$$\begin{array}{ccc} A & \longrightarrow & C \\ \downarrow & & \downarrow \\ B & \longrightarrow & D \end{array}$$

of CW complexes (where all of the maps are cellular inclusions) be taken to a pullback square of groups

$$\begin{array}{ccc} H_*A & \longrightarrow & H_*C \\ \downarrow & & \downarrow \\ H_*B & \longrightarrow & H_*D. \end{array}$$

By setting C and D to be the cone on A, and $D = \Sigma A$, the isomorphism (1.0.1) follows from a simple calculation.

The functor Q satisfies a related homotopical condition: that any *homotopy* pushout square of spaces

$$X \longrightarrow W$$
$$\downarrow \qquad \downarrow$$
$$Y \longrightarrow Z$$

be taken to a *homotopy* pullback square

$$QX \longrightarrow QW$$
$$\downarrow \qquad\quad \downarrow$$
$$QY \longrightarrow QZ.$$

Such functors are called *excisive*, or more specifically 1-*excisive*. In [43], Goodwilllie shows that any functor $F\colon \mathsf{Top}_* \to \mathsf{Top}_*$ which preserves weak homotopy equivalences admits a universal such 1-excisive approximation, $P_1 F$, by means of a comparison map $F \to P_1 F$ (which, under suitable niceness conditions on F should induce an isomorphism in a "stable range" of homotopy groups when evaluated on suitably connected spaces). Indeed, the universal 1-excisive approximation to the identity functor on Top_* is the stable homotopy functor Q.

Let Spt denote (a model for) the category of *spectra* symmetric monoidal with respect to the smash product $\wedge$, for which the sphere spectrum S is the unit[3]. The term *linear* can be justified by observing that any 1-excisive functor $\mathsf{Spt} \to \mathsf{Spt}$ is of the form $X \mapsto E \wedge X$ for fixed spectrum E[4]. A 1-excisive functor $F\colon \mathsf{Top}_* \to \mathsf{Top}_*$ such that $F(*) \simeq *$ will similarly always factor through the the usual *stabilization adjunction* $(\Sigma^\infty, \Omega^\infty)$ between Top_* and Spt,

$$\mathsf{Top}_* \underset{\Omega^\infty}{\overset{\Sigma^\infty}{\rightleftarrows}} \mathsf{Spt}\,,$$

as $F \simeq \Omega^\infty \widetilde{F} \Sigma^\infty$ such that $\widetilde{F}$ is a 1-excisive functor on spectra. In particular, the universal

---

[3]Such as the symmetric spectra of [51], or S-modules of [36].
[4]At least when restricted to *finite* spectra X.

1-excisive approximation is the usual *stabilization* of F,

$$P_1 F(X) \simeq \mathrm{colim}_n \, \Omega^n F(\Sigma^n X)$$

and the spectrum which classifies $\widetilde{P_1 F}$ is called the *first derivative* of F [41].

The Taylor tower

In fact, Goodwillie shows more. In [43], he constructs a *Taylor tower* of n-excisive or "polynomial of degree at most n" functors, $P_n F$, along with and natural transformations $P_{n+1}F \to P_n F$ (for $n \geq 0$), of the form

$$F \to \cdots \to P_n F \to P_{n-1}F \to \cdots \to P_1 F \to P_0 F.$$

These functors $P_n F$ play a role analogous to the Taylor polynomials $p_n f$ associated to an infinitely differentiable function $f \colon \mathbf{R} \to \mathbf{R}$ in ordinary calculus and analysis. The formula

$$p_n f(x) - p_{n-1}f(x) = \frac{f^{(n)}(0)}{n!}x^n$$

has a homotopical analog in that the "difference" (in this case, *homotopy fiber*) between $P_n F$ and $P_{n-1}F$ is measured by a certain spectrum with action by the n-th symmetric group Σ_n denoted $\partial_n F$ (the *n-th derivative* of F) as an equivalence of the form

$$D_n F(X) := \mathrm{hofib}(P_n F(X) \to P_{n-1}F(X)) \simeq \Omega^\infty (\partial_n F \wedge \Sigma^\infty X^{\wedge n})_{h\Sigma_n}.$$

The subscript $h\Sigma_n$ above denotes that homotopy orbits have been taken with respect to the Σ_n action on $\partial_n F$ and on $\Sigma^\infty X^{\wedge n}$ (via permutation of factors).

Under suitable niceness conditions on F and the input space X, the Taylor tower may be used to tell us something about the homotopy type of $F(X)$. For instance, the Taylor tower of the identity on Top_* will converge to the homotopy type of its input when evaluated on simply-connected spaces. That is, for 1-connected $X \in \mathsf{Top}_*$, the natural map

$$X \to \lim_{n \to \infty} P_n \mathsf{Id}_{\mathsf{Top}_*}(X)$$

is a weak homotopy equivalence. As such, the Taylor tower of the identity provides a useful canonical resolution of a space which begins with the stabilization QX. Moreover, this resolution is "controlled" by the symmetric sequence of derivatives $\partial_* \mathsf{Id}_{\mathsf{Top}_*} = \{\partial_n \mathsf{Id}_{\mathsf{Top}_*}\}_{n \geq 1}$

Much work has gone into understanding Taylor towers and Goodwillie derivatives in recent decades, both for functors of spaces and spectra and also in more general contexts (see, for instance, [66], [11], [60, §6]). Johnson [52] and Arone-Mahowald [7] give a description of the layers $D_n \mathsf{Id}_{\mathsf{Top}_*}$ in terms of Spanier-Whitehead duals to the n-th *partition poset complex*—essentially, a pointed space built from the combinatorics of partitions of the set $\{1, \ldots, n\}$. For instance, the 2-nd partition poset complex is just S^0 with trivial Σ_2 action, and so

$$D_2 \mathsf{Id}_{\mathsf{Top}_*}(X) \simeq \Omega^\infty \mathrm{Map}(S^0, \Sigma^\infty X^{\wedge 2})_{\Sigma_2} \simeq Q\Omega X^{\wedge 2}_{\Sigma_2}.$$

Similarly, $\partial_2 \mathsf{Id}_{\mathsf{Top}_*} \simeq \Omega S$, a desuspension of the sphere spectrum, with trivial Σ_2 action.

Spectral Lie algebras

Using this description, Ching [22] has shown that the symmetric sequence of Goodwillie derivatives $\partial_* \mathsf{Id}_{\mathsf{Top}_*} = \{\partial_n \mathsf{Id}_{\mathsf{Top}_*}\}_{n \geq 1}$ can be given a natural *operad* structure. In fact, this operad fits nicely into a broader story of *Koszul duality* as described by [40], [39], [38]. The partition poset models described by [52], [7] for $\partial_* \mathsf{Id}_{\mathsf{Top}_*}$ indeed show that the derivatives of the identity in spaces are Koszul dual to the commutative *cooperad*[5] in Spt, and as such is often referred to as the *Spectral Lie operad* (as the Lie operad in chain complexes in similarly

---

[5]See Appendix B for our treatment of cooperads

dual to the commutative cooperad [40]). This spectral Lie operad additionally plays central rule in describing a *chain rule* for derivatives as shown by Arone-Ching [2]: that is, an equivalence of symmetric sequences

$$\partial_* F \circ_{\partial_* \mathrm{Id}_{\mathsf{Top}_*}} \partial_* G \simeq \partial_*(FG)$$

for functors $F, G \colon \mathsf{Top}_* \to \mathsf{Top}_*$ (here $\circ$ denotes the *composition product*, see 2.1.6).

More generally, the operad of Goodwillie derivatives plays a crucial role in describing the homotopy theory of the category of based spaces. Heuts [50] (see also [12]) has further shown that certain "chromatic localizations" M_n^f of Top_* can be characterized by the category of algebras over $\partial_* \mathrm{Id}_{\mathsf{Top}_*}$ in $T(n)$-local spectra. As such, it is often anticipated that similar techniques may be used to better understand other categories in which one can *do* functor calculus. A widely recognized slogan of functor calculus that the Goodwillie derivatives of the identity on a suitable model category C should come equipped with a canonical operad structure which tells us something about C. One such example, and the focus of our first main theorem, is when C is a category of algebras over a (reduced) operad $\mathcal{O}$ in spectra.

Operads and structured ring spectra

First, some background: Operads [62], [17] are combinatorial tools for describing spaces (or spectra) X which admit a pairing $X \wedge X \to X$ and unit map $S \to X$ that may be only associative or commutative up to coherent homotopies (such as the structure found on an n-fold loop space $\Omega^n Y$ [62]) and have played an increasingly common and powerful role in homotopy theory. Common examples of interest are associative, commutative, or E_n-ring spectra (for $1 \leq n \leq \infty$)—the latter of which interpret between *highly homotopy associative* (E_1 or A_∞) and *highly homotopy commutative* (E_∞) ring spectra and enjoy a rich structure on the (co)homology theories they represent. For an operad $\mathcal{O}$ in spectra, we write $\mathsf{Alg}_\mathcal{O}$ for the category of algebras over $\mathcal{O}$. An object $X \in \mathsf{Alg}_\mathcal{O}$ is a spectrum X together with associative and unital action maps of the form

$$\mathcal{O}[n] \wedge_{\Sigma_n} X^{\wedge n} \to X \qquad (n \geq 0)$$

We will require that our operads be *reduced* in that the 0-ary operations described by $\mathcal{O}$ are trivial (i.e., $\mathcal{O}[0] = *$). Such condition guarantees that our $\mathcal{O}$-algebras are *nonunital*; a condition that can naturally arises when working with augmented rings.

Pereira [66] shows that the constructions and main results of [43] readily extend to functors between categories of algebras over operads in spectra. Many other authors are responsible for early work which suggests this should be possible including Kuhn [57], McCarthy-Minasian [64], Basterra-Mandell [9], and Harper-Hess [46]. Our work relies heavily on the constructions of Kuhn-Pereira [55] for describing the Taylor tower of certain functors on $\mathsf{Alg}_\mathcal{O}$.

1.1 Main theorems

Our first main result, Theorem 1.1.1(a) as follows, is that the Goodwillie derivatives of the identity in the category of algebras over a reduced operad $\mathcal{O}$ in spectra can be given a naturally occurring "highly homotopy coherent" operad structure.

Theorem 1.1.1. *Let $\mathcal{O}$ be an operad in spectra such that $\mathcal{O}[n]$ is (-1)-connected for $n \geq 1$ and $\mathcal{O}[0] = *$. Then,*

(a) *The derivatives of the identity in $\mathsf{Alg}_\mathcal{O}$ can be equipped with a natural highly homotopy coherent operad structure.*

(b) *Moreover, with respect to this structure, $\partial_* \mathrm{Id}_{\mathsf{Alg}_\mathcal{O}}$ is equivalent to $\mathcal{O}$ as highly homotopy coherent operads.*

It has been known for some time that $\mathcal{O}[n]$ is a model for $\partial_n \mathrm{Id}_{\mathsf{Alg}_\mathcal{O}}$. Part (b) of Theorem 1.1.1 in particular answers a long standing conjecture which appears in [2] (and also answered in the context of ∞-operads by Ching in [26]), a main difficulty of which is describing an intrinsic operad structure on the derivatives of the identity which may be compared with that of the operad $\mathcal{O}$.

The proofs of parts (a) and (b) to Theorem 1.1.1 may be found in Sections 6.1.1 and 6.1.3, respectively. Our technique is to avoid working with the identity directly by replacing it with the Bousfield-Kan cosimplicial resolution provided by the stabilization adjunction (Q, U) for $\mathcal{O}$-algebras. The strong cartesianness estimates of Blomquist [14] (see also Ching-Harper [27]) allow us to then express $\partial_* \mathrm{Id}_{\mathsf{Alg}_\mathcal{O}}$ as the homotopy limit of the cosimplicial diagram (showing only coface maps)

$$\partial_*(QU)^{\bullet+1} = \left(\partial_*(UQ) \rightrightarrows \partial_*(UQ)^2 \mathrel{\substack{\longrightarrow\\[-0.6ex]\longrightarrow\\[-0.6ex]\longrightarrow}} \partial_*(UQ)^3 \cdots \right) \tag{1.1.2}$$

whose terms $\partial_*(QU)^{k+1}$ may be readily computed by an $\mathcal{O}$-algebra analogue of the Snaith splitting. We thus obtain a natural cosimplicial resolution $C(\mathcal{O})$ of the derivatives of the identity such that $\partial_* \mathrm{Id}_{\mathsf{Alg}_\mathcal{O}} \simeq \operatorname{holim}_\Delta C(\mathcal{O})$ which furthermore may be identified as the TQ resolution of $\mathcal{O}$ as a left $\mathcal{O}$-module. Our approach is influenced by the work of Arone-Kankaanrinta [6] wherein they use the cosimplicial resolution offered by the stabilization adjunction between spaces and spectra to analyze the derivatives of the identity in spaces via the classic Snaith splitting.

We induce a highly homotopy coherent operad structure (i.e., A_∞-operad) on $\partial_* \mathrm{Id}_{\mathsf{Alg}_\mathcal{O}}$ by constructing a pairing of the resolution $C(\mathcal{O})$ with respect to the box product $\square$ for cosimplicial objects (see Batanin [10]). Thus, we extend to the monoidal category of symmetric sequences a technique utilized in McClure-Smith [65]: specifically, that if X is a $\square$-monoid in cosimplicial spaces or spectra then $\mathsf{Tot}(X)$ is an A_∞-monoid (with respect to the closed, symmetric monoidal product for spaces or spectra).

There are some subtleties that arise in that (i) the box product is not as well-behaved when working with the composition product $\circ$ of symmetric sequences, and (ii) the extra structure encoded by $\circ$ leads us to work with $\mathbf{N}$-colored operads to express A_∞-monoids with respect to composition product. As such, one of the main developments of this **book** is that of $\mathbf{N}$-colored operads *with levels* (i.e., $\mathbf{N}_{\text{lev}}$-operads) as useful bookkeeping tools designed to algebraically encode operads (i.e., strict composition product monoids) and "fattened-up" operads as their algebras. Within this framework of $\mathbf{N}_{\text{lev}}$-operads we can also describe algebras over an A_∞-operad.

Remark 1.1.3. In the statement of Theorem 1.1.1 the phrase "naturally occuring" means that we refrain from endowing $\partial_* \mathrm{Id}_{\mathsf{Alg}_\mathcal{O}}$ with the operad structure from $\mathcal{O}$ directly. Rather, we produce a method for intrinsically describing operadic structure possessed by the derivatives of the identity that should carry over to other model categories suitable for functor calculus. In particular, the constructions of such an operad structure on the derivatives of the identity should:

(i) Recover the (A_∞-) operad structure endowed on $\partial_* \mathrm{Id}_{\mathsf{Top}_*}$ described by Ching in [22].

(ii) Endow the derivatives of an suitably nice homotopy functor[6] $F \colon \mathsf{Alg}_\mathcal{O} \to \mathsf{Alg}_{\mathcal{O}'}$ with a natural $(\partial_* \mathrm{Id}_{\mathsf{Alg}_{\mathcal{O}'}}, \partial_* \mathrm{Id}_{\mathsf{Alg}_\mathcal{O}})$-bimodule structure (which is in turn equivalent to an $(\mathcal{O}', \mathcal{O})$-bimodule) suitable for describing a chain rule (as in Arone-Ching [2]) with a view toward running the machinery of [3].

(iii) Be fundamental enough to describe an operad structure on $\partial_* \mathrm{Id}_\mathsf{C}$ and chain rule for a suitable model category C (e.g., one in which one can do functor calculus).

In fact, item (i) is the subject of our second main result, Theorem 1.1.4 below. Items (ii) and (iii) are the subject of current work, and we defer a discussion to our progress along with some conjectural remarks to Section 6.3.

---

[6]For instance, we should expect this to be possible for functors which are finitary and simplicial

Theorem 1.1.4. *The symmetric sequence $\partial_* \mathrm{Id}_{\mathsf{Top}_*}$ is a highly homotopy coherent operad.*

One of our main interests in operads is in what structure is described on their algebras. Ching further shows that the homology of $\partial_* \mathrm{Id}_{\mathsf{Top}_*}$ gives a "Lie operad" [22], which is *Koszul dual* to the commutative cooperad [40] (see also [39]). The operad $\partial_* \mathrm{Id}_{\mathsf{Top}_*}$ of spectra is similarly Koszul dual to the commutative cooperad in spectra (as in [38]) and so is often referred to as the *spectral Lie operad*.

In addition, we prove the following theorem (Theorem 1.1.5). As the commutative cooperad in Spt is also an operad, Theorem 1.1.5 may be thought of as an amplified version of the classic result from commutative algebra which states that the primitives of a Hopf algebra naturally form a Lie algebra (see, e.g. [1]).

Theorem 1.1.5. *The derived primitives (Definition B.3.5) of a commutative coalgebra[7] in spectra admit a natural action by $\partial_* \mathrm{Id}_{\mathsf{Top}_*}$.*

The above theorem is also not necessarily new. For a finite commutative coalgebra Y, the derived primitives $\mathsf{Prim}(Y)$ is equivalent to $\mathsf{TQ}(Y^\vee)^\vee$ (see Example B.3.6)—where here $Y^\vee = \mathrm{Map}(Y, S)$ is the Spanier-Whitehead dual—and the latter is known to admit an action by $\partial_* \mathrm{Id}_{\mathsf{Top}_*}$ (e.g. as in [22], [12], [50]). However, our constructions provide an easy to use, algebraic alternative to the pioneering operad structure on $\partial_* \mathrm{Id}_{\mathsf{Top}_*}$ found in [22]. In particular, we expect our approach to allow us to show that the Goodwillie derivatives of the identity in a suitable model category C come equipped with a canonical operad structure, induced in a similar way (see 6.3.6).

1.1.6 Outline of the argument

Our main tool is to utilize the Bousfield-Kan cosimplicial resolution with respect to the stabilization. For the context of $\mathcal{O}$-algebras in Theorem 1.1.1, this relies on a result of

---

[7]For us, coalgebra means *coalgebra with divided powers* (see, e.g. [38]).

Basterra [8], that the stabilization of an $\mathcal{O}$-algebra X is naturally equivalent to its *topological Quillen homology* spectrum $\mathsf{TQ}(X)$ (see Section 2.3). Topological Quillen homology may be thought of as the derived indecomposables quotient for structured ring spectra described as algebras over a (reduced) operad, and arises as the left derived functor of the left adjoint in

$$\mathsf{Alg}_{\mathcal{O}} \underset{U}{\overset{Q}{\rightleftarrows}} \mathsf{Alg}_J \simeq \mathsf{Mod}_{\mathcal{O}[1]}.$$

Here, J denotes a suitable replacement of $\tau_1\mathcal{O}$, the truncation of $\mathcal{O}$ above level 1 (see Section 2.3), U denotes the forgetful functor along the map of operads $\mathcal{O} \to J$ (which also provides trivial operations from $\mathcal{O}[n]$ on a J-algebra for $n \geq 2$), and Q is the $\mathcal{O}$-algebra analog of the indecomposables quotient R/R^2 for R a nonunital commutative ring in classical algebra.

Using the strong connectivity estimates offered by Blomquist's higher stabilization theorems [14, §7], we first show that $\partial_*\mathrm{Id}_{\mathsf{Alg}_{\mathcal{O}}}$ is equivalent to $\mathrm{holim}_{\Delta}\,\partial_*(UQ)^{\bullet+1}$ (see (1.1.2)). Similarly to Arone-Kankaanrinta [6], in which they compute the n-excisive approximations (resp. n-th derivatives) of the identity functor on Top_* in terms of the n-excisive approximations (resp. n-th derivatives) of iterates of stabilization $\Omega^\infty\Sigma^\infty$ by means of the Snaith splitting, we then analyze the terms $\partial_*(UQ)^{k+1}$ via an analog of the Snaith splitting in $\mathsf{Alg}_{\mathcal{O}}$.

Essentially a statement about the Taylor tower of the associated comonad QU, the Snaith splitting in $\mathsf{Alg}_{\mathcal{O}}$ permits equivalences of symmetric sequences

$$\partial_*(QU) \simeq |\operatorname{Bar}(J, \mathcal{O}, J)| \simeq J \circ_{\mathcal{O}}^{\mathsf{h}} J =: B(\mathcal{O})$$

as (J, J)-bimodules (here, $\circ^{\mathsf{h}}$ denotes the derived composition product). By iterated applications of the splitting, we may compute

$$\partial_*(UQ)^{k+1} \simeq \underbrace{B(\mathcal{O}) \circ_J \cdots \circ_J B(\mathcal{O})}_{k} \simeq \underbrace{J \circ_{\mathcal{O}} \cdots \circ_{\mathcal{O}} J}_{k+1} = C(\mathcal{O})^k$$

and moreover that $\partial_*(UQ)^{\bullet+1} \simeq C(\mathcal{O})$ as cosimplicial symmetric sequences. Here, $C(\mathcal{O})$ is given by

$$J \rightrightarrows J \circ_{\mathcal{O}} J \Rrightarrow J \circ_{\mathcal{O}} J \circ_{\mathcal{O}} J \Rrightarrow J \circ_{\mathcal{O}} J \circ_{\mathcal{O}} J \circ_{\mathcal{O}} J \cdots$$

with coface map d^i induced by inserting $\mathcal{O} \to J$ at the i-th position (see Remark 3.3.11 along with (2.3.4)).

Note, $B(\mathcal{O})$ is (at least up to homotopy) a cooperad with a coaugmentation map $J \to B(\mathcal{O})$, and our $C(\mathcal{O})$ is essentially a rigid cosimplicial model for the cobar construction on $B(\mathcal{O})$. In particular, this allows us to bypass referencing any particular model for the comultiplication on $B(\mathcal{O})$ (e.g., that of Ching [22], see also Section 3.3.4).

We construct a pairing $m \colon C(\mathcal{O}) \square C(\mathcal{O}) \to C(\mathcal{O})$ with respect to the box product (Definition 4.1.1) of cosimplicial symmetric sequences via compatible maps of the form (induced by the operad structure maps $J \circ J \to J$)

$$m_{p,q} \colon \underbrace{J \circ_{\mathcal{O}} \cdots \circ_{\mathcal{O}} (J \circ J) \circ_{\mathcal{O}} \cdots \circ_{\mathcal{O}} J}_{p+1 \qquad\qquad q+1} \to \underbrace{J \circ_{\mathcal{O}} \cdots \circ_{\mathcal{O}} J \circ_{\mathcal{O}} \cdots \circ_{\mathcal{O}} J}_{p+q+1}$$

along with a unit map $u \colon \underline{I} \to C(\mathcal{O})$, where $\underline{I}$ denotes the constant cosimplicial symmetric sequence on I. Our argument is then to induce an A_∞-monoidal pairing on $\partial_* \mathrm{Id}_{\mathsf{Alg}_{\mathcal{O}}}$ — modeled as $\mathsf{Tot}\, C(\mathcal{O})$ — via m and u (compare with McClure-Smith [65]).

One difficulty which arises is that the composition product of symmetric sequences is not as well-behaved of a product as, say, cartesian product of spaces or smash product of spectra. Thus, we *do not* obtain m as a strictly monoidal pairing on the level of cosimplicial diagrams. In resolving this issue we introduce a specialized category of $\mathbf{N}$-colored operads *with levels* (i.e., $\mathbf{N}_{\mathsf{lev}}$-operads) designed specifically to overcome these technical subtleties of the composition product. As a result, a large portion of this document is dedicated to carefully developing the framework of $\mathbf{N}_{\mathsf{lev}}$-operads and their algebras.

With these details in tow it is then possible to produce an A_∞-operad structure on

$\partial_* \mathrm{Id}_{\mathsf{Alg}_\mathcal{O}}$. Let Tot denote restricted totalization $\mathsf{Tot}^{\mathsf{res}}$ (see Section 2.2.2), we then obtain an A_∞-monoidal pairing

$$\mathsf{Tot}\, C(\mathcal{O}) \circ \mathsf{Tot}\, C(\mathcal{O}) \to \mathsf{Tot}\, C(\mathcal{O})$$

described as an algebra over a certain $\mathbf{N}_{\mathsf{lev}}$-operad which is a naturally "fattened-up" replacement of the $\mathbf{N}_{\mathsf{lev}}$-operad whose algebras are strict operads (see Definition 5.4.2 along with Propositions 5.4.5 and 5.4.23). Moreover, the coaugmentation $\mathcal{O} \to C(\mathcal{O})$ provides a comparison between $\mathcal{O}$ and $\partial_* \mathrm{Id}_{\mathsf{Alg}_\mathcal{O}}$ which we show yields an equivalence of A_∞-operads, thus resolving the aforementioned conjecture.

Our method for proving Theorems 1.1.4 and 1.1.5 is similar. We make use of the models for $\partial_n \mathrm{Id}_{\mathsf{Top}_*}$ as the Spanier-Whitehead dual of the n-th partition poset complex $\mathsf{Par}(n)$ (Definition 3.4.5),

$$\partial_n \mathrm{Id}_{\mathsf{Top}_*} \simeq \mathrm{Map}(\mathsf{Par}(n), S), \qquad (n \geq 1).$$

We then show that $\mathrm{Map}(\mathsf{Par}(n), S)$ can be modeled as the totalization of the cobar resolution $C(\underline{S})[n]$ on the commutative cooperad of spectra. Essentially, $C(\underline{S})[n]$ has p-simplices given by the p-fold composition $\underline{S}^{\circ p}[n]$, where $\underline{S}$ is the reduced symmetric sequence with $\underline{S}[n] = S$ (with trivial Σ_n action) for $n \geq 1$. We show that the cosimplicial symmetric sequence $C(\underline{S})$ admits a natural $\mathring{\square}$-monoid structure (see [29, Definition 5.6]) induced, essentially, by the canonical isomorphisms $\underline{S}^{\circ p} \circ \underline{S}^{\circ q} \cong \underline{S}^{\circ p+q}$. This $\mathring{\square}$-monoid structure then induces the desired "highly homotopy coherent" operad structure on $\partial_* \mathrm{Id}_{\mathsf{Top}_*}$ upon passing to totalization.

We similarly show that the derived primitives may be calculated as the totalization of a certain cosimplicial spectrum $C(Y)$. In particular, $C(Y)$ is precisely the dual notion of derived indecomposables which underlies the construction of topological Quillen homology for $\mathcal{O}$-algebras. We show that $C(Y)$ is a $\mathring{\square}$-module over the cobar resolution $C(\underline{S})$ and that this module structure induces an action of $\partial_* \mathrm{Id}_{\mathsf{Top}_*}$ on the derived primitives upon passing to totalization.

1.1.7 Comparison to the operad structure constructed by Ching

We expect the operad structure on $\partial_* \mathrm{Id}_{\mathsf{Top}_*}$ described in this document is equivalent to that constructed by Ching in [22]. In particular, our main technical lemma (Lemma 6.2.1) is reminiscent of the "tree ungrafting" arguments found in [22], and our notion of $\mathring{\square}$-monoid is similar to Ching's notion of "pre-cooperad" in [24]. However, despite these similarities, the author is not aware of an explicit comparison between the two constructions, nor to the third description offered in [26].

1.1.8 Organization of this document

This document is organized into seven Chapters and features two appendices A and B.

Chapter 2 is devoted to the required background for working with operads and their algebras in spectra as well as stabilization and topological Quillen homology of structured ring spectra.

Chapter 3 gives a brief overview of Goodwillie's calculus of homotopy functors and also describes the our main objects of interest, specifically, the Goodwillie derivatives of the identity functor.

Chapter 4 introduces the box product for cosimplicial diagrams and offers some technical remarks for extending this product to the category of cosimplicial symmetric sequences.

Chapter 5 introduces our notion of $\mathbf{N}$-colored operads with levels (i.e. $\mathbf{N}_{\mathsf{lev}}$-operads) and describes a model for "fattened-up operads" which we will employ.

Chapter 6 contains proofs of our main theorems on the deriavtives of the identity in structured ring spectra and in the category of pointed topological spaces, and also includes some conjectural remarks on extending our techniques to a more general setting.

Appendix A contains a deferred proof of a particular model for the n-excisive approximations to certain functors of structured ring spectra. Appendix B contains the necessary background for (a particular model of) cooperads and their coalgebras.

Chapter 2

Operads and their algebras

We work in the category algebras over a reduced operad in a closed, symmetric monoidal category of spectra $(\mathsf{Spt}, \wedge, S)$. For convenience we will use the category of S-modules as in Elmendorf-Kriz-Mandell-May [36] and refer to such objects as *spectra*. The main technical benefit of working with S-modules is that all spectra will be fibrant (and thus $\mathrm{Tot}\,\mathcal{X}$ of a levelwise fibrant diagram will already correctly model $\mathrm{holim}_\Delta \mathcal{X}$), though we note that similar results should hold in the category of symmetric spectra by utilizing suitable fibrant replacement monads.

We observe that Spt is a cofibrantly generated, closed symmetric monoidal model category (see, e.g., [2, Definition 1.12]) and write $\mathrm{Map}^{\mathsf{Spt}}(X, Y)$ for the internal mapping object of Spt. When it is clear from context we write Map for $\mathrm{Map}^{\mathsf{Spt}}$. We let Top denote the category of compactly generated Hausdorff spaces. In [36], it is shown that Spt admits a tensoring of Top_* which may be extended to Top by first adding a disjoint basepoint. In particular for $K \in \mathsf{Top}$, $X, Y \in \mathsf{Spt}$ there are natural isomorphisms

$$\hom(K_+ \wedge X, Y) \cong \hom(X, \mathrm{Map}^{\mathsf{Spt}}(K_+, Y)).$$

Though we will not make explicit use of it, we define a simplicial tensoring of Spt via

$K \wedge X := |K| \wedge X$ for $K \in \mathsf{sSet}_*$ and $X \in \mathsf{Spt}$.

2.1 Symmetric sequences

Let $(\mathsf{C}, \otimes, \mathbf{1})$ be a closed symmetric monoidal category and write $\mathrm{Map}^{\mathsf{C}}$ for the mapping object in C. When C is clear from context we write Map for $\mathrm{Map}^{\mathsf{C}}$. We will require that C be cocomplete, and write ϕ for the initial object of C; particular categories of interest are Spt and Top_*.

Recall that a *symmetric sequence* in C is a collection $X[n] \in \mathsf{C}$ for $n \geq 0$ such that $X[n]$ admits a (right) action by Σ_n. We let $\mathsf{SymSeq}_{\mathsf{C}}$ denote the category of symmetric sequences in C and action preserving morphisms. A symmetric sequence X is *reduced* if $X[0] = \phi$ (some authors require in addition that $X[1] \cong \mathbf{1}$, however we omit this condition). When C is clear from context we will simply write SymSeq. Note that SymSeq comes equipped with a monoidal product $\circ$, the *composition product* (also called *circle product*) defined as follows (see also [67] or [44]).

2.1.1 The composition product of symmetric sequences

For $X, Y \in \mathsf{SymSeq}$ we define $X \circ Y$ at level k by

$$(X \circ Y)[k] = \coprod_{n \geq 0} X[n] \otimes_{\Sigma_n} Y^{\check{\otimes} n}[k]. \tag{2.1.2}$$

Here, $\check{\otimes}$ denotes the *tensor* of the symmetric sequences (e.g., as in [44]). For $n, k \geq 0$, $Y^{\check{\otimes} n}[k]$ is computed as

$$\coprod_{k \xrightarrow{\pi} n} Y[\pi_1] \otimes \cdots \otimes Y[\pi_n] \cong \coprod_{k_1 + \cdots + k_n = k} \Sigma_k \cdot_{\Sigma_{k_1} \times \cdots \times \Sigma_{k_n}} Y[k_1] \otimes \cdots \otimes Y[k_n]$$

where π runs over all surjections $\mathbf{k} = \{1, \ldots, n\} \to \{1, \ldots, n\} = \mathbf{n}$ and we set $\pi_i := |\pi^{-1}(i)|$ for $i \in \mathbf{n}$. The composition product admits a unit I given by $I[1] = \mathbf{1}$ and $I[k] = \phi$ otherwise.

For our purposes, we find it convenient to work with with a slightly modified version of the composition product for reduced symmetric sequences. Let $X, Y \in \mathsf{SymSeq}$ be reduced. Let $(k_1, \ldots, k_n)$ denote a sequence of integers $k_1, \ldots, k_n \geq 1$ (allowing for repetition of entries) and set Sum_n^k to be the collection of orbits $(k_1, \ldots, k_n)_{\Sigma_n}$ such that $\sum_{i=1}^n k_i = k$.

Definition 2.1.3. Given $k_1, \ldots, k_n \geq 1$ we define $H(k_1, \ldots, k_n)$ as the collection of block permutation matrices $\Sigma_{k_1} \times \cdots \times \Sigma_{k_n} \leq \Sigma_k$, along with the Σ_{p_i} permutations of those blocks such that $k_j = d_i$.

Remark 2.1.4. We observe that orbits $(k_1, \ldots, k_n)_{\Sigma_n}$ are in bijective correspondence to partitions $k = d_1 p_1 + \cdots + d_m p_m$ where $1 \leq d_1 < \cdots < d_m$ and $p_i \geq 1$. Given an orbit $(k_1, \ldots, k_n)_{\Sigma_n}$ let $1 \leq d_1 \leq \cdots d_m$ be the distinct entries of multiplicity p_i. We note that there is an isomorphism (here, $\Sigma_m^{\wr n}$ denotes the *wreath product* $\Sigma_m \wr \Sigma_n := \Sigma_m^{\times n} \rtimes \Sigma_n$)

$$H(k_1, \ldots, k_n) \cong \Sigma_{d_1}^{\wr p_1} \times \cdots \times \Sigma_{d_m}^{\wr p_m}.$$

Moreover $H(k_1, \ldots, k_n)$ admits a natural Σ_n action by permutation of elements k_i and the induced map $H(k_1, \ldots, k_n) \to H(k_{\sigma(1)}, \ldots, k_{\sigma(n)})$ is an isomorphism for all $\sigma \in \Sigma_n$.

Though we will not need this fact, we remark that $H(k_1, \ldots, k_n)$ may be identified with the stabilizer of the Σ_k action on partitions of $\{1, \ldots, k\}$ into sets of size $k_1, \ldots, k_n$ (see, e.g., [23, §1.12]).

For $k \geq 0$ we set $\Sigma[k] := \coprod_{\sigma \in \Sigma_k} \mathbf{1}$.

Remark 2.1.5. The composition product $X \circ Y$ may be equivalently written as

$$(X \circ Y)[k] \cong \coprod_{n \geq 0} \coprod_{(k_1, \ldots, k_n)_{\Sigma_n} \in \mathsf{Sum}_n^k} \Sigma[k] \otimes_{H(k_1, \ldots, k_n)} X[n] \otimes Y[k_1] \otimes \cdots \otimes Y[k_n]. \tag{2.1.6}$$

Here, the action of $H(k_1, \ldots, k_n)$ on $\Sigma[k]$ is induced by that on Σ_k and the action on $X[n] \otimes Y[k_1] \otimes \cdots \otimes Y[k_n]$ is given as follows (see also Ching [23, 1.13])

- $\Sigma_{p_1} \times \cdots \times \Sigma_{p_m} \leq \Sigma_n$ acts on $X[n]$

- for $i = 1, \ldots, m$, $\Sigma_{d_i}^{\wr p_i}$ acts on the factors $Y[k_j]$ such that $k_j = d_i$ by

  (i) permuting the p_i factors $Y[d_i]$

  (ii) acting by corresponding Σ_{d_i} factor on each $Y[d_i]$.

We also make the following definition for the *nonsymmetric composition product* $X \hat{\circ} Y$ (note that our definition differs from [45])

$$(X \hat{\circ} Y)[k] := \coprod_{n \geq 0} \coprod_{(k_1, \ldots, k_n)_{\Sigma_n} \in \mathsf{Sum}_n^k} X[n] \otimes Y[k_1] \otimes \cdots \otimes Y[k_n]. \tag{2.1.7}$$

Note that $\hat{\circ}$ is not associative, our primary use for $\hat{\circ}$ will be as a bookkeeping tool for indexing the factors involved in expanding iterates of $\circ$ from the left (as in Section 5).

2.1.8 Operads as monads

An *operad* in C is a symmetric sequence $\mathcal{O}$ which is a monoid with respect to $\circ$, i.e., there are maps $\mathcal{O} \circ \mathcal{O} \to \mathcal{O}$ and $I \to \mathcal{O}$ which satisfy additional associativity and unitality relations (see, e.g., Rezk [67]). An operad is *reduced* if $\mathcal{O}[0] = *$. We will only consider reduced operads in this document, and interpret *operad* to mean reduced operad unless otherwise specified.

Any symmetric sequence M gives rise to a functor $M \circ (-)$ on C given as follows (note $X^{\otimes 0} = 1$)

$$X \mapsto M \circ (X) = \bigvee_{n \geq 0} M[n] \otimes_{\Sigma_n} X^{\otimes n}.$$

If $\mathcal{O}$ is an operad, then the associated functor $\mathcal{O} \circ (-)$ is a monad on C which we will frequently conflate with the operad $\mathcal{O}$. We let $\mathsf{Alg}_{\mathcal{O}}^{\mathsf{C}}$ denote the category of algebras for the monad associated to an operad $\mathcal{O}$ in C.

2.2   Some assumptions and notation for structured ring spectra

When $\mathsf{C} = \mathsf{Spt}$ and $\mathcal{O}$ is an operad of spectra, then $\mathsf{Alg}_{\mathcal{O}} = \mathsf{Alg}_{\mathcal{O}}^{\mathsf{Spt}}$ is a pointed simplicial model category (see, e.g., [28, §7]) when endowed with projective model structure from Spt. For a further overview of notation and terminology we refer the reader to [44, §3] or [67, §2].

2.2.1   Assumptions on $\mathcal{O}$

From now on in this document we assume that $\mathcal{O}$ is a reduced operad in Spt which obeys some mild cofibrancy conditions that are satisfied if, e.g., $\mathcal{O}$ arises via the suspension spectra of a cofibrant operad in spaces. In particular, we require that the underlying symmetric sequence of $\mathcal{O}[n]$ be Σ-cofibrant (see, e.g., [2, §9]) and that the terms $\mathcal{O}[n]$ be (-1)-connected for all $n \geq 1$.

2.2.2   Use of restricted totalization

We systematically interpret Tot of a cosimplicial diagram to mean restricted totalization (see also [28, §8])

$$\mathsf{Tot} := \mathrm{Tot}^{\mathrm{res}} \cong \mathrm{Map}_{\boldsymbol{\Delta}^{\mathrm{res}}}^{\mathsf{Spt}} \left(\Delta^{\bullet}, - \right) := \left(\mathrm{Map}^{\mathsf{Spt}}(\Delta^{\bullet}, -) \right)^{\boldsymbol{\Delta}^{\mathrm{res}}}.$$

Here, $\boldsymbol{\Delta}$ denote the usual simplicial category of finite totally-ordered sets $[n] := \{0 < 1 < \cdots < n\}$ and order preserving maps, $\boldsymbol{\Delta}^{\mathrm{res}} \subset \boldsymbol{\Delta}$ is the subcategory obtained by omitting degeneracy maps, and $\Delta^{\bullet}$ denotes the usual cosimplicial space of topological n-simplices. For convenience, if $C^\bullet$ is cosimplicial object, we will write $\mathsf{Tot}\, C^\bullet$ instead of the more technically correct $\mathsf{Tot}(C^\bullet|_{\Delta^{\mathrm{res}}})$.

Diagrams shaped on Δ^{res} are referred to as *restricted* cosimplicial diagrams. Importantly the inclusion $\Delta^{\mathrm{res}} \to \Delta$ is homotopy left cofinal[8] and so if $C^\bullet$ is a cosimplicial diagram in $\mathsf{Alg}_{\mathcal{O}}$ which is levelwise fibrant (as opposed to the stronger condition of Reedy fibrancy), there are equivalences

$$\mathrm{holim}_{\Delta}\, C^\bullet \simeq \mathrm{holim}_{\Delta^{\mathrm{res}}}\, C^\bullet \simeq \mathsf{Tot}\, C^\bullet.$$

2.2.3 Truncations of $\mathcal{O}$

For $n \geq 0$ we define $\tau_n \colon \mathsf{SymSeq} \to \mathsf{SymSeq}$ to be the *n-th truncation functor* given at a symmetric sequence M by

$$(\tau_n M)[k] = \begin{cases} M[k] & k \leq n \\ * & k > n \end{cases}$$

with natural transformations $\tau_n \to \tau_{n-1}$. We let i_n be the fiber of $\tau_n \to \tau_{n-1}$, i.e., $i_n M[k] = *$ for $k \neq n$ and $i_n M[n] = M[n]$ in which case we say $i_n M$ is *concentrated at level n*.

For $M = \mathcal{O}$ the truncations $\tau_n \mathcal{O}$ assemble into a tower of $(\mathcal{O}, \mathcal{O})$-bimodules which receives a map from $\mathcal{O}$ of the form

$$\mathcal{O} \to \cdots \to \tau_3\mathcal{O} \to \tau_2\mathcal{O} \to \tau_1\mathcal{O}. \tag{2.2.4}$$

The tower (2.2.4) is well studied and plays a central role in examining the homotopy completion of a structured ring spectrum as in [46]. Note as well that $\mathcal{O} \to \tau_1\mathcal{O}$ is a map of operads and there is a well-defined composite $\tau_1\mathcal{O} \to \mathcal{O} \to \tau_1\mathcal{O}$ which factors the identity on $\tau_1\mathcal{O}$.

---

[8]The main property we are interested in here is that such functors induce equivalences on homotopy limits.

2.2.5 Change of operad adjunction

Associated to a map $f\colon \mathcal{O} \to \mathcal{O}'$ of operads there is a Quillen adjunction of the form (see, e.g., [67])

$$\mathsf{Alg}_{\mathcal{O}} \underset{f^*}{\overset{f_*}{\rightleftarrows}} \mathsf{Alg}_{\mathcal{O}'}$$

in which the left adjoint f_* is given by the (reflective) coequalizer

$$f_*(X) := \mathcal{O}' \circ_{\mathcal{O}} (X) = \operatorname{colim} \left(\mathcal{O}' \circ \mathcal{O} \circ (X) \rightrightarrows \mathcal{O}' \circ (X) \right)$$

and the right adjoint f^* is the forgetful functor along f. If f is a levelwise equivalence then the above adjunction is a Quillen equivalence and furthermore the left derived functor $\mathsf{L}f_*$ may be calculated via a simplicial bar construction as follows (see, e.g., [44])

$$\mathsf{L}f_*(X) := \mathcal{O}' \circ_{\mathcal{O}}^{\mathsf{h}} (X) \simeq |\operatorname{Bar}(\mathcal{O}', \mathcal{O}, X^c)|.$$

2.3 Stabilization of structured ring spectra

In order to have a well-defined calculus of functors on $\mathsf{Alg}_{\mathcal{O}}$ it is necessary to understand the stabilization of the category of such algebras. Note that $\mathsf{Alg}_{\mathcal{O}}$ is tensored over simplicial sets (see, e.g., [28, §7]) and thus one can define $\mathsf{Sp}(\mathsf{Alg}_{\mathcal{O}})$, the category of Bousfield-Friendlander spectra of $\mathcal{O}$-algebras, which is Quillen equivalent to the category of left $\mathcal{O}[1]$-modules, $\mathsf{Mod}_{\mathcal{O}[1]}$ (see, e.g., [9] or [66, §2]).

The stabilization map for $\mathcal{O}$-algebras is thus equivalent to the left adjoint of (2.2.5) with respect to the map of operads $\mathcal{O} \to \tau_1 \mathcal{O}$, i.e.,

$$\Sigma^{\infty}_{\mathsf{Alg}_{\mathcal{O}}} X \simeq \tau_1 \mathcal{O} \circ_{\mathcal{O}} (X)$$

for $\mathcal{O}$-algebras X. By analogy, $\Omega^\infty_{\mathsf{Alg}_\mathcal{O}}$ gives an $\mathcal{O}[1]$-module trivial $\mathcal{O}$-algebra structure above level 2. Moreover, if $\mathcal{O}[1] \cong S$, then the stabilization of $\mathsf{Alg}_\mathcal{O}$ is equivalent to the underlying category Spt.

As in [28], we replace $\tau_1\mathcal{O}$ by a "fattened-up" operad J to produce an iterable model for TQ-homology with the right homotopy type. That is, let J be any factorization

$$\mathcal{O} \xrightarrow{h} J \xrightarrow{g} \tau_1\mathcal{O}$$

in the category of operads, where h is a cofibration and g a weak equivalence. There are then change of operads adjunctions

$$\mathsf{Alg}_\mathcal{O} \underset{U}{\overset{Q}{\rightleftarrows}} \mathsf{Alg}_J \underset{g^*}{\overset{g_*}{\rightleftarrows}} \mathsf{Alg}_{\tau_1\mathcal{O}} \cong \mathsf{Mod}_{\mathcal{O}[1]} \tag{2.3.1}$$

such that (g_*, g^*) is a Quillen equivalence and, notably, U preserves cofibrant objects (see [46, 5.49]). We refer to the pair (Q, U) as the stabilization adjunction for $\mathcal{O}$-algebras and use Alg_J as our model for the stabilization of $\mathsf{Alg}_\mathcal{O}$.

2.3.2 TQ-homology

The total left derived functor $\mathsf{LQ}(X) =: \mathsf{TQ}(X)$ is called the **TQ**-*homology spectrum* of X and the composite $\mathsf{R}U(\mathsf{LQ}(X))$ is the **TQ**-*homology* $\mathcal{O}$-*algebra* of X. We note that the TQ-homology spectrum of X may be calculated in the underlying category Spt as

$$\mathsf{LQ}(X) \simeq |\,\mathrm{Bar}(J, \mathcal{O}, X^c)| \simeq |\,\mathrm{Bar}(\tau_1\mathcal{O}, \mathcal{O}, X^c)|.$$

For simplicity, we will assume the $\mathcal{O}$-algebras we work with are cofibrant by first replacing X by X^c, where $(-)^c$ denotes a functorial cofibrant replacement in $\mathsf{Alg}_\mathcal{O}$.

2.3.3 The Bousfield-Kan resolution with respect to TQ

Associated to the stabilization adjunction for $\mathcal{O}$-algebras (Q, U) there is a comonad $\mathsf{K} := QU$ on Alg_J. Given Y a K-coalgebra, we let $C(Y)$ denote the cosimplicial object $\mathrm{Cobar}(U, \mathsf{K}, Y)$.

For $X \in \mathsf{Alg}_{\mathcal{O}}$, let $X \to C(X) := C(QX)$ be the coaugmented cosimplicial object given below

$$X \to \Big(UQ(X) \rightrightarrows (UQ)^2(X) \substack{\longrightarrow \\ \longrightarrow \\ \longrightarrow} (UQ)^3(X) \cdots \Big) \tag{2.3.4}$$
$$\cong \Big(J \circ_{\mathcal{O}} (X) \rightrightarrows J \circ_{\mathcal{O}} J \circ_{\mathcal{O}} (X) \substack{\longrightarrow \\ \longrightarrow \\ \longrightarrow} J \circ_{\mathcal{O}} J \circ_{\mathcal{O}} J \circ_{\mathcal{O}} (X) \cdots \Big)$$

Coface maps d^i in (2.3.4) are induced by inserting $\mathcal{O} \to J$ at the i-th position, i.e.,

$$J \circ_{\mathcal{O}} \cdots \circ_{\mathcal{O}} J \cong J \circ_{\mathcal{O}} \cdots \circ_{\mathcal{O}} \mathcal{O} \circ_{\mathcal{O}} \cdots \circ_{\mathcal{O}} J \to J \circ_{\mathcal{O}} \cdots \circ_{\mathcal{O}} J \circ_{\mathcal{O}} \cdots \circ_{\mathcal{O}} J$$

and codegeneracy maps s^j are induced by $J \circ_{\mathcal{O}} J \to J \circ_J J \cong J$ at the j-th position.

Remark 2.3.5. The totalization of the diagram (2.3.4) above is called the TQ-completion of an $\mathcal{O}$-algebra X, defined by

$$X^{\wedge}_{\mathsf{TQ}} := \mathsf{Tot}\, C(X) \simeq \mathrm{holim}_{\Delta}\, C(X).$$

It is known that $X \simeq X^{\wedge}_{\mathsf{TQ}}$ for any 0-connected $\mathcal{O}$-algebra X (see, e.g., [27]).

2.3.6 Cubical diagrams

Let $\mathcal{P}(n)$ denote the poset of subsets of the set $\{1, \ldots, n\}$. A functor $\mathcal{Z} \colon \mathcal{P}(n) \to \mathsf{C}$ is called an *n-cube in* C or also an *n-cubical diagram*. We use the following notation $\mathcal{P}_0(n) := \mathcal{P}(n) \setminus \{\emptyset\}$ and $\mathcal{P}_1(n) := \mathcal{P}(n) \setminus \{\{1, \ldots, n\}\}$ and refer to diagrams shaped on either $\mathcal{P}_0(n)$ or $\mathcal{P}_1(n)$ as *punctured n-cubes*. The *total homotopy fiber* of an n-cube $\mathcal{Z}$, denoted tohofib $\mathcal{Z}$, is defined to be the homotopy fiber of the natural comparison map

$$\chi_0 \colon \mathcal{Z}(\emptyset) \to \operatorname{holim}_{\mathcal{P}_0(n)} \mathcal{Z}.$$

If the comparison χ_0 is a weak equivalence (resp. k-connected) we say that $\mathcal{Z}$ is *homotopy cartesian* (resp. k-cartesian).

Dually, the *total homotopy cofiber* of $\mathcal{Z}$ is the homotopy cofiber of the comparison map

$$\chi_1 \colon \operatorname{hocolim}_{\mathcal{P}_1(n)} \mathcal{Z} \to \mathcal{Z}(\{1, \ldots, n\})$$

which we denote by tohocofib $\mathcal{Z}$. If χ_1 is a weak equivalence (resp. k-connected) we say that $\mathcal{Z}$ is *homotopy cocartesian* (resp. k-cocartesian). We note that the total homotopy fiber (resp. cofiber) of a cube may be calculated by iterated homotopy fibers (resp. cofibers), see e.g., [11, 3.2].

Example 2.3.7 (Coface n-cube). Let $\mathcal{Z}^{-1} \xrightarrow{d^0} \mathcal{Z}^\bullet$ be a coaugmented cosimplicial object. There are associated coface n-cubes $\mathcal{Z}_n$ whose subfaces encode the relation on coface maps (see, e.g., Ching-Harper [28, §2.3]). We demonstrate $\mathcal{Z}_2$ and $\mathcal{Z}_3$ below

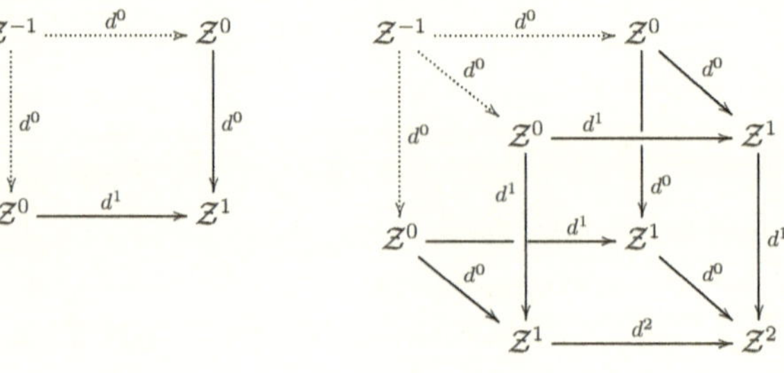

2.3.8 Higher stabilization for $\mathcal{O}$-algebras

For $k \geq 0$, let $\mathbf{\Delta}^{\leq k}$ denote the full subcategory of $\mathbf{\Delta}$ comprised of sets $[\ell] \in \mathbf{\Delta}$ for $\ell \leq k$ (note $\mathbf{\Delta}^{\leq -1} = \emptyset$). There are inclusions of categories

$$\emptyset = \mathbf{\Delta}^{\leq -1} \to \mathbf{\Delta}^{\leq 0} \to \mathbf{\Delta}^{\leq 1} \to \cdots \to \mathbf{\Delta}^{\leq k} \to \cdots \to \mathbf{\Delta}$$

and moreover $\mathrm{holim}_{\mathbf{\Delta}} Y$ may be computed as limit of the tower $\{\mathrm{holim}_{\mathbf{\Delta}^{\leq k}} Y\}$ (see, e.g., [28, §8.11] for a detailed write-up). There is a natural homotopy left cofinal inclusion $\mathcal{P}_0(n) \to \mathbf{\Delta}^{\leq n-1}$ which, in particular, allows us to model the comparison $X \to \mathrm{holim}_{\mathbf{\Delta}^{\leq n-1}} C(X)$ via the map χ_0 (see Section 2.3.6) for the coface n-cube associated to $X \to C(X)$.

By careful examination of the connectivities of these maps, Blomquist is able to obtain the following strong convergence estimates as a corollary to [14, 7.1] (see also Dundas [33] and Dundas-Goodwillie-McCarthy [34]).

Proposition 2.3.9. *Let $\mathcal{O}$ be an operad in* Spt *whose entries are* (-1)*-connected, $X \in \mathsf{Alg}_{\mathcal{O}}$ k-connected, and $C(X)$ as in (2.3.5). Then, for any $n \geq 0$ the induced map $X \to \mathrm{holim}_{\mathbf{\Delta}^{\leq n-1}} C(X)$ is $(k+1)(n+1)$-connected.*

These estimates show, in particular, if X is 0-connected then $X \xrightarrow{\sim} X^{\wedge}_{\mathsf{TQ}}$ (see also Ching-Harper [27]).

Chapter 3

An overview of functor calculus

Functor calculus was introduced by Goodwillie in a landmark series of papers [41, 42, 43] as a means of analyzing homotopy functors to or from Top_* or Spt. Since, the theory been recognized as a general phenomenon which, in particular, relates a suitable model category to its stabilization. We refer the reader to [5] for an overview and exposition of some recent applications of the theory.

In what follows we consider functors $F\colon \mathsf{C} \to \mathsf{D}$ between suitable pointed simplicial model categories C and D. Our main examples are when C, D are either Top_*, Spt or when C, D are categories of structured ring spectra described as algebras over reduced operads in spectra. We refer the reader to Pereira [66] for a more detail on functor calculus in categories of structured ring spectra.

3.1 The Taylor tower

Definition 3.1.1. A functor $F\colon \mathsf{C} \to \mathsf{D}$ is n-excisive if F takes any strongly cocartesian $(n+1)$-cube in C to a cartesian $(n+1)$-cube in D.

Note, when $n = 1$ this is precisely the excisive condition discussed in the introduction, as a strongly cocartesian 2-cube is just a homotopy pushout cube.

Example 3.1.2. One nugget of intuition for the above definition is that n-excisive functors are "determined" by their values on $n+1$ points, similar to polynomial functions $f\colon \mathbf{R} \to \mathbf{R}$, as follows. Let $X \in \mathsf{Top}_*$ and let $\mathcal{X}$ be the strongly cocartesian $(n+1)$-cube in Top_* with $\mathcal{X}(\emptyset) = X$ and $\mathcal{X}(\{k\}) = CX$, the cone on X with $\mathcal{X}(\emptyset) \to \mathcal{X}(\{k\})$ the usual inclusion, for $1 \leq k \leq n+1$. A functor $F\colon \mathsf{Top}_* \to \mathsf{Top}_*$ being n-excisive in particular asks that $F(\mathcal{X})$ be a cartesian $(n+1)$-cube in Top_*. Unravelling definitions, this tells us that $F(X) = F(\mathcal{X}(\emptyset))$ be recovered by the homotopy limit of the cube $F(\mathcal{X}|\mathcal{P}_0(n+1))$; which is in turn determined by the values of $F(*) \simeq F(CX) = F(\mathcal{X}(\{k\}))$ for $1 \leq k \leq n+1$.

A central construction in functor calculus is that of the *Taylor tower* (sometimes referred to also as the *Goodwillie tower*) of n-excisive approximations associated to a functor $F\colon \mathsf{C} \to \mathsf{D}$ as follows

$$
\begin{array}{c}
D_n F \\
\downarrow \\
F \longrightarrow \cdots \longrightarrow P_n F \longrightarrow P_{n-1} F \longrightarrow \cdots \longrightarrow P_0 F.
\end{array}
\tag{3.1.3}
$$

The functor $P_n F$ is called the *n-th excisive approximation* to F and is initial in the homotopy category of n-excisive functors receiving a map from F. In this work, all of our approximations are based at the zero object $* \in \mathsf{C}$, though approximations based at arbitrary spaces $Y \in \mathsf{Top}_*$ are described in [43]. The functor $D_n F$ is called the *n-th homogeneous layer* and is defined as

$$
D_n F := \mathrm{hofib}(P_n F \to P_{n-1} F).
$$

Note that $P_0 F$ is a constant functor taking value $F(*)$. We call F *reduced* if $F(*) \simeq *$ and note that for reduced functors we have $P_1 F \simeq D_1 F$. Recipes for constructing $P_n F$ are provided in [43] for functors of Top_* and Spt, and [66] for functors of structured ring spectra.

3.1.4 Analytic functors

If F satisfies additional connectivity conditions on certain cubical diagrams (e.g., if F is suitably stably n-excisive for all n as in [42, 4.1]) we call F *analytic*, or more specifically *ρ-analytic*: a key feature being that an analytic functor F may be recovered as the homotopy limit of the tower (3.1.3) on ρ-connected inputs X, i.e.,

$$F(X) \simeq \mathrm{holim}_n\, P_n F(X).$$

For instance, the identity functor on Top_* is 1-analytic by the higher Blakers-Massey theorems (see, e.g., [42, §2]) and the analogous results for structured ring spectra of Ching-Harper [27] demonstrate that the identity functor on $\mathsf{Alg}_{\mathcal{O}}$ is 0-analytic.

3.1.5 Cross effects and derivatives

Let $\mathcal{S}_n(X_1, \ldots, X_n)$ denote the n-cube

$$T \mapsto \bigvee_{t \notin T} X_t, \text{ for } T \in \mathcal{P}(n).$$

The *n-th cross effect* of G is the n-variable functor defined by

$$\mathrm{cr}_n G(X_1, \ldots, X_n) := \mathrm{tohofib}\, G(\mathcal{S}_n(X_1, \ldots, X_n)).$$

Our work concerns the derivatives of a functor F, which are certain spectra which classify the homogeneous layers $D_n F$ (under some mild conditions on F) and are computable via cross effects. We recall first that a functor G is *n-homogeneous* if G is n-excisive and $P_k G \simeq *$ for $k < n$ and that G is *finitary* if G commutes with filtered homotopy colimits.

A major triumph of functor calculus is the classification of n-homogeneous functors.

Proposition 3.1.6 below is summarized from Goodwillie [43] (for functors of spaces and spectra) and highlights the relevant properties of the homogeneous layers $D_n F$ and derivatives $\partial_n F$ associated to a functor F.

Proposition 3.1.6. *Let $F\colon \mathsf{Top}_* \to \mathsf{Top}_*$ be a homotopy functor and $n \geq 1$. Then:*

- $D_n F$ *is n-homogeneous.*

- $D_n F$ *naturally factors through* Spt *as* $D_n F \simeq \Omega^\infty \circ \mathbb{D}_n F \circ \Sigma^\infty$ *such that* $\mathbb{D}_n F$ *is n-homogeneous.*

- $D_n F$ *is characterized by a spectrum with (right) Σ_n-action, $\partial_n F$ called the n-th derivative of F, and there is an equivalence*[9]

$$D_n F(X) \simeq \Omega^\infty(\partial_n F \wedge_{\Sigma_n} (\Sigma^\infty X)^{\wedge n}).$$

- *The spectrum $\partial_n F$ may be calculated via cross effects [43, §3] as*

$$\partial_n F \simeq \mathrm{cr}_n \mathbb{D}_n F(S, \ldots, S)$$

with Σ_n-action given by permutting the inputs S.

3.2 Functor calculus in categories of structured ring spectra

We have a similar proposition for functors of structured ring spectra, summarized from [66]. For notational convenience we let $\widetilde{\mathsf{TQ}}$ denote the composite $g_* \mathsf{TQ} \simeq \tau_1 \mathcal{O} \circ_{\mathcal{O}}^{\mathsf{h}} (-)$.

---

[9]For X which have the homotopy type of a finite CW-complexes; or on an arbitrary space X if F commutes with filtered colimits (i.e., F is *finitary*).

Proposition 3.2.1. *Let* $F\colon \mathsf{Alg}_{\mathcal{O}} \to \mathsf{Alg}_{\mathcal{O}}$ *be a homotopy functor,* $X \in \mathsf{Alg}_{\mathcal{O}}$, *and* $n \geq 1$. *Then:*

(i) $D_n F$ *is* n-*homogeneous.*

(ii) *There are* n-*homogeneous functors* $\mathbb{D}_n F$ *and* $\widetilde{\mathbb{D}_n} F$ *such that the following diagram commutes*

$$
\begin{array}{ccccc}
\mathsf{Alg}_{\mathcal{O}} & \xrightarrow{\;Q\;} & \mathsf{Alg}_J & \xrightarrow{\;g_*\;} & \mathsf{Mod}_{\mathcal{O}[1]} \\
\Big\downarrow{\scriptstyle D_n F} & & \Big\downarrow{\scriptstyle \mathbb{D}_n F} & & \Big\downarrow{\scriptstyle \widetilde{\mathbb{D}_n} F} \\
\mathsf{Alg}_{\mathcal{O}} & \xleftarrow{\;U\;} & \mathsf{Alg}_J & \xleftarrow{\;g^*\;} & \mathsf{Mod}_{\mathcal{O}[1]}
\end{array}
\tag{3.2.2}
$$

(iii) *There is a* (J, J)-*bimodule* $\partial_* F$, *whose* n-*th entry* $\partial_n F$ *is called the* n-*th Goodwillie derivative of* F, *and such that there are equivalences of underlying spectra*

$$
D_n F(X) \simeq i_n(\partial_* F) \circ_J^{\mathsf{h}} (\mathsf{TQ}(X))
$$

(iv) $D_n F$ *is characterized by an* $(\mathcal{O}[1], \Sigma_n \wr \mathcal{O}[1])$-*bimodule*[10] $\widetilde{\partial_n} F$ *which has underlying spectrum equivalent to that of* $\partial_n F$.

(v) *There are equivalences of underlying spectra*

$$
D_n F(X) \simeq (\widetilde{\partial_n} F \wedge^{\mathsf{L}}_{\mathcal{O}[1]^{\wedge n}} \widetilde{\mathsf{TQ}}(X)^{\wedge^{\mathsf{L}} n})_{h\Sigma_n} \simeq \widetilde{\partial_n} F \wedge^{\mathsf{L}}_{\Sigma_n \wr \mathcal{O}[1]} \widetilde{\mathsf{TQ}}(X)^{\wedge^{\mathsf{L}} n}.
\tag{3.2.3}
$$

(vi) *The* n-*th derivative may be calculated via* n-*th cross effects* cr_n *as*

$$
\partial_n F \simeq \widetilde{\partial_n} F \simeq \mathrm{cr}_n \widetilde{\mathbb{D}_n} F(\mathcal{O}[1], \ldots, \mathcal{O}[1])
$$

with right $\Sigma_n \wr \mathcal{O}[1]$-*action granted by permuting the inputs.*

---

[10]That is, a left module over $\mathcal{O}[1]$ and right module over $\Sigma_n \wr \mathcal{O}[1]$ (see Definition 3.2.7 for the definition of the wreath product $\Sigma_n \wr \mathcal{O}[1]$)

Remark 3.2.4. The above equivalence (3.2.3) hold in general for finite cell $\mathcal{O}$-algebras X and, if F further is *finitary* (i.e., F commutes with filtered homotopy colimits), then the equivalences may be extended to arbitrary $\mathcal{O}$-algebras X. The notation $\wedge^{\mathsf{L}}$ and $\circ^{\mathsf{h}}$ denote the derived smash product and circle product, respectively. We will often omit the latter notation and understand our constructions to be implicitly derived.

The careful reader might note that the n-th Goodwillie derivative of F is only defined up to weak equivalence, and so the choice $\partial_n F$ vs. $\widetilde{\partial}_n F$ for functors of structured ring spectra may seem a pedantic distinction. For our purposes, this distinction is beneficial to the readibility of several of the upcoming proofs. Further, there are equivalences

$$\mathsf{L}g_* \partial_n F \simeq \widetilde{\partial}_n F \text{ and } \partial_n F \simeq \mathsf{R}g^* \widetilde{\partial}_n F,$$

and for concreteness, the model for the derivatives of the identity we employ is as a (J, J)-bimodule, $\mathsf{Tot}\, C(\mathcal{O})$ (see (3.3.12)).

Of note is that the choice of $\mathbb{D}_n F$ (resp. $\widetilde{\mathbb{D}}_n F$) may be made functorial in F by a straightforward modification of the argument presented in [2, 2.7]. In particular if F is finitary, then for any $Y \in \mathsf{Mod}_{\mathcal{O}[1]}$ we have

$$\widetilde{\mathbb{D}}_n F(Y) \simeq \widetilde{\partial}_n F \wedge_{\Sigma_n \wr \mathcal{O}[1]} Y^{\wedge n}. \tag{3.2.5}$$

3.2.6   A note on wreath products

We use $\Sigma_n \wr \mathcal{O}[1]$ to denote the twisted group ring (i.e., *wreath product*) $(\Sigma_n)_+ \wedge \mathcal{O}[1]^{\wedge n}$. We recall some pertinent details of wreath products of ring spectra below.

Definition 3.2.7. Given a ring spectrum $\mathcal{R}$ we define

$$\Sigma_n \wr \mathcal{R} := \Sigma_n \cdot \mathcal{R}^{\wedge n} \cong (\Sigma_n)_+ \wedge \mathcal{R}^{\wedge n}$$

with multiplication given by

$$(\sigma \wedge x) \wedge (\tau \wedge y) \mapsto \sigma\tau \wedge x\sigma(y).$$

Our main use of such objects stems from the following proposition (see also [58, Lemma 14], [55, §2]). Note that a right $\Sigma_n \wr \mathcal{R}$-module is a (right) Σ_n object via the unit map $I \to \mathcal{R}$.

Proposition 3.2.8. *Let $\mathcal{R}$ be a ring spectrum, X a left $\mathcal{R}$-module and M a right $\mathcal{R}$-module with n commuting actions of $\mathcal{R}$ (i.e., right $\mathcal{R}^{\wedge n}$-module). Then, there is an isomorphism*

$$(M \wedge_{\mathcal{R}^{\wedge n}} X^{\wedge n})_{\Sigma_n} \cong M \wedge_{\Sigma_n \wr \mathcal{R}} X^{\wedge n}.$$

Remark 3.2.9. The right-hand equivalence of (3.2.3) is an instance of this equivalence. Of note is that if X is a cofibrant $\mathcal{O}$-algebra, then $\mathsf{TQ}(X)$ is cofibrant in $\mathsf{Mod}_{\mathcal{O}[1]}$ and therefore Proposition 3.2.8 provides that $\mathsf{TQ}(X)^{\wedge n}$ is a cofibrant as a left $\Sigma_n \wr \mathcal{O}[1]$-module.

In addition, the $(\mathcal{O}[1], \Sigma_n \wr \mathcal{O}[1])$-bimodule structure on the derivatives $\widetilde{\partial}_n F$ for all $n \geq 1$ induces $(\tau_1\mathcal{O}, \tau_1\mathcal{O})$-bimodule structure on the symmetric sequence $\widetilde{\partial}_* F$ which is further compatible with the (J, J)-bimodule structure on $\partial_* F$ via the (g_*, g^*) adjunction. In the simplified case that $\mathcal{O}[1] \cong S$, an $(S, \Sigma_n \wr S)$-bimodule is just a spectrum with a right action by Σ_n and (3.2.3) reduces to an equivalence of underlying spectra

$$D_n F(X) \simeq \widetilde{\partial}_n F \wedge_{\Sigma_n} \widetilde{\mathsf{TQ}}(X)^{\wedge n}.$$

3.2.10 Taylor towers of certain functors $\mathsf{Alg}_{\mathcal{O}} \to \mathsf{Alg}_{\mathcal{O}'}$

Let $\mathcal{O}$, $\mathcal{O}'$ be reduced operads in spectra and M be a cofibrant $(\mathcal{O}', \mathcal{O})$-bimodule with $M[0] = *$ whose terms are (-1)-connected. We define a functor $F_M \colon \mathsf{Alg}_{\mathcal{O}} \to \mathsf{Alg}_{\mathcal{O}'}$ at

$X \in \mathsf{Alg}_{\mathcal{O}}$ by the simplicial bar construction

$$F_M(X) = |\operatorname{Bar}(M, \mathcal{O}, X)| \simeq M \circ^h_{\mathcal{O}} (X). \tag{3.2.11}$$

Note F_M is finitary and the left $\mathcal{O}'$ action on M induces a left $\mathcal{O}'$ action on $F_M(X)$. The following proposition may be summarized from Harper-Hess [46] and Kuhn-Pereira [55, §2.7] and further provides a model for the Taylor tower of functors F_M. For completion, we sketch proofs of the relevant details.

Proposition 3.2.12. *Let M and F_M be as described above. Then there are equivalences (natural in M)*

(i) $P_n F_M \simeq \tau_n M \circ^h_{\mathcal{O}} (-)$

(ii) $D_n F_M \simeq i_n M \circ^h_{\mathcal{O}} (-) \simeq i_n M \circ^h_J (\mathsf{TQ}(-))$

(iii) $\widetilde{\mathbb{D}_n} F_M(-) \simeq M[n] \wedge^L_{\Sigma_n \wr \mathcal{O}[1]} (-)^{\wedge n}$

(iv) $\widetilde{\partial_n} F_M \simeq M[n]$

such that the Taylor tower for F_M is equivalent to

$$
\begin{array}{c}
i_n M \circ^h_{\mathcal{O}} (-) \\
\downarrow \\
F_M \longrightarrow \cdots \longrightarrow \tau_n M \circ^h_{\mathcal{O}} (-) \longrightarrow \tau_{n-1} M \circ^h_{\mathcal{O}} (-) \longrightarrow \cdots \longrightarrow \tau_1 M \circ^h_{\mathcal{O}} (-).
\end{array}
$$

Proof. We will write $\circ$ for $\circ^h$ and $\wedge$ for $\wedge^L$. The equivalence (i) is proved in Appendix A (see also [66, 4.3]). For (ii) we note that morphisms $\tau_n M \to \tau_{n-1} M$ give rise to the comparison maps on excisive approximations $P_n F_M \xrightarrow{q_n} P_{n-1} F_M$ and moreover the fiber sequence

$$i_n M \to \tau_n M \to \tau_{n-1} M$$

identifies $i_n M \circ_{\mathcal{O}} (-)$ with the fiber of q_n. Moreover, as the right $\mathcal{O}$-action on $i_n M$ factors through $\tau_1 \mathcal{O}$ there are then equivalences of underlying spectra

$$D_n F_M(X) \simeq (i_n M \circ_{\tau_1 \mathcal{O}} \tau_1 \mathcal{O}) \circ_{\mathcal{O}} (X)$$

$$\simeq i_n M \circ_{\tau_1 \mathcal{O}} (\tau_1 \circ_{\mathcal{O}} (X)) \simeq i_n M \circ_J (\mathsf{TQ}(X)).$$

Note that (iii) follows from the observation that any $Y \in \mathsf{Mod}_{\mathcal{O}[1]}$

$$i_n M \circ_{\tau_1 \mathcal{O}} (Y) \simeq M[n] \wedge_{\Sigma_n \wr \mathcal{O}[1]} Y^{\wedge n}.$$

The proof of (iv) follows from the equivalence $\mathrm{cr}_n F \simeq \mathrm{cr}^n F$ between cross-effects and *co-cross-effects* of functors landing in a stable category as in Ching [23] (see also McCarthy [63]), where latter is defined dually to cr_n as follows

$$\mathrm{cr}^n G(X_1, \ldots, X_n) = \mathrm{tohocofib}\left(\mathcal{P}(n) \ni T \mapsto G\left(\prod_{t \in T} X_t \right) \right).$$

In particular, taking co-cross-effects will commute with $\wedge_{\Sigma_n \wr \mathcal{O}[1]}$ and so

$$\mathrm{cr}_n \widetilde{\mathbb{D}_n} F_M \simeq \mathrm{cr}_n(M[n] \wedge_{\Sigma_n \wr \mathcal{O}[1]} (-)^{\wedge n}) \simeq M[n] \wedge_{\Sigma_n \wr \mathcal{O}[1]} \mathrm{cr}_n((-)^{\wedge n}).$$

Via the computation $\mathrm{cr}_n((-)^{\wedge n}) \simeq (\Sigma_n)_+ \wedge (-)^{\wedge n}$ we then obtain

$$\widetilde{\partial_n} F_M \simeq M[n] \wedge_{\Sigma_n \wr \mathcal{O}[1]} (\Sigma_n \wr \mathcal{O}[1]) \simeq M[n].$$

$\square$

Definition 3.2.13. For functors of the form F_M we take as our models for $P_n F_M$, $D_n F_M$ and $\widetilde{\partial_n} F_M$ those from Proposition 3.2.12. A map $M \to M'$ of cofibrant $(\mathcal{O}, \mathcal{O})$-bimodules induces natural transformations $P_n F_M \to P_n F_{M'}$ and $D_n F_M \to D_n F_{M'}$, and also that $\widetilde{\partial_n} F_M \to \widetilde{\partial_n} F_{M'}$

is equivalent to $M[n] \to M'[n]$.

3.2.14 The Taylor tower of the identity on $\mathsf{Alg}_{\mathcal{O}}$

Note that for $M = \mathcal{O}$, the functor $F_{\mathcal{O}}$ is equivalent to the identity via $\mathcal{O} \circ_{\mathcal{O}} (-) \simeq \mathrm{Id}_{\mathsf{Alg}_{\mathcal{O}}}$. Moreover, there are natural transformations $\mathrm{Id}_{\mathsf{Alg}_{\mathcal{O}}} \to \tau_n \mathcal{O} \circ_{\mathcal{O}} (-)$ provided by the unit map of the change of operads adjunction (Section 2.2.5) applied to the map of operads $\mathcal{O} \to \tau_n \mathcal{O}$. The Taylor tower of the identity in $\mathsf{Alg}_{\mathcal{O}}$ then is equivalent to

$$
\begin{array}{c}
i_n \mathcal{O} \circ_{\mathcal{O}} (-) \\
\downarrow \\
\mathrm{Id}_{\mathsf{Alg}_{\mathcal{O}}} \longrightarrow \cdots \longrightarrow \tau_n \mathcal{O} \circ_{\mathcal{O}} (-) \longrightarrow \tau_{n-1} \mathcal{O} \circ_{\mathcal{O}} (-) \longrightarrow \cdots \longrightarrow \tau_1 \mathcal{O} \circ_{\mathcal{O}} (-)
\end{array}
\tag{3.2.15}
$$

This tower (3.2.15) has previously been studied by Harper-Hess [46] in relation to homotopy completion of $\mathcal{O}$-algebras (see also Kuhn [56] and McCarthy-Minasian [64]). Moreover, Ching-Harper provide $\mathsf{Alg}_{\mathcal{O}}$ analogues of the higher Blakers-Massey theorems in [27] which in particular show that $\mathrm{Id}_{\mathsf{Alg}_{\mathcal{O}}}$ is 0-analytic. That is, for 0-connected X the following comparison map is an equivalence

$$
X \to \mathrm{holim}_n \, \tau_n \mathcal{O} \circ_{\mathcal{O}} (X).
$$

As a corollary to Proposition 3.2.12, we obtain equivalences of underlying spectra (see also [46])

$$
D_n \mathrm{Id}_{\mathsf{Alg}_{\mathcal{O}}}(X) \simeq i_n \mathcal{O} \circ_{\mathcal{O}} (X) \simeq \mathcal{O}[n] \wedge_{\Sigma_n \wr \mathcal{O}[1]} \widetilde{\mathsf{TQ}}(X)^{\wedge n}
$$

and also observe that $\widetilde{\partial}_n \mathrm{Id}_{\mathsf{Alg}_{\mathcal{O}}} \simeq \mathcal{O}[n]$ as a $(\mathcal{O}[1], \Sigma_n \wr \mathcal{O}[1])$-bimodule for all $n \geq 1$. Therefore, with a view toward the operad structure on $\partial_* \mathrm{Id}_{\mathsf{Top}_*}$ constructed by Ching in [22] we are lead to the following question, found in Arone-Ching [2]: *Is it possible to endow $\partial_* \mathrm{Id}_{\mathsf{Alg}_{\mathcal{O}}}$ with a naturally occurring operad structure such that $\partial_* \mathrm{Id}_{\mathsf{Alg}_{\mathcal{O}}} \simeq \mathcal{O}$ as operads?*

A key idea to our approach is taken from Arone-Kankaanrinta [6] where they show that $\partial_* \mathrm{Id}_{\mathsf{Top}_*}$ may be better understood by utilizing the cosimplicial resolution from the stabilization adjunction $(\Sigma^\infty, \Omega^\infty)$ by means of the Snaith splitting. Within the realm of 0-connected $\mathcal{O}$-algebras, the (Q, U) adjunction between $\mathsf{Alg}_{\mathcal{O}}$ and Alg_J (the latter, recall, is Quillen equivalent to $\mathsf{Mod}_{\mathcal{O}[1]}$) is the exact analogue of stabilization. We provide an $\mathsf{Alg}_{\mathcal{O}}$ analogue of the Snaith splitting in Section 3.3.1.

3.3   A model for derivatives of the identity in $\mathsf{Alg}_{\mathcal{O}}$

The aim of this section is to describe specifically the model for the derivatives of the identity we employ, as Tot of a certain cosimplicial symmetric sequence $C(\mathcal{O})$ which may be motivated as the totalization of the cosimplicial object arising from a calculation of the n-th derivative of $(QU)^k$ via the Snaith splitting in $\mathsf{Alg}_{\mathcal{O}}$. We are further motivated by work of Arone-Kankaanrinta [6] which utilizes the Snaith splitting in spaces (3.4.12) to provide a model for the derivatives of the identity in spaces.

3.3.1   The Snaith splitting in $\mathsf{Alg}_{\mathcal{O}}$

There is an analogous result for $\mathcal{O}$-algebras, wherein the adjunction $(\Sigma^\infty, \Omega^\infty)$ is replaced by (Q, U) from (2.3.1). Let $B(\mathcal{O})$ be the (J, J)-bimodule

$$B(\mathcal{O}) = J \circ_{\mathcal{O}}^{\mathsf{h}} J \simeq |\operatorname{Bar}(J, \mathcal{O}, J)|$$

and note that given $Y \in \mathsf{Alg}_J$ cofibrant there is a zig-zag of equivalences.

$$QU(Y) \xleftarrow{\sim} |\operatorname{Bar}(J, \mathcal{O}, Y)| \xrightarrow{\sim} |\operatorname{Bar}(J, \mathcal{O}, J)| \circ_J (Y) = B(\mathcal{O}) \circ_J (Y).$$

The $\mathsf{Alg}_{\mathcal{O}}$ *Snaith splitting* is then the equivalence

$$QU(Y) \simeq B(\mathcal{O}) \circ_J (Y).\tag{3.3.2}$$

Remark 3.3.3. At first blush, (3.3.2) may not seem like a proper "splitting" in the style of the classic Snaith splitting for Top_* (see Section 3.4.12), which declares an equivalence

$$\Sigma^{\infty}\Omega^{\infty}\Sigma^{\infty}X \simeq \bigvee_{k \geq 1} \Sigma^{\infty} X_{h\Sigma_k}^{\wedge k}$$

for $X \in \mathsf{Top}_*$. This is more an artifact of our use of Alg_J for the stabilization of $\mathsf{Alg}_{\mathcal{O}}$. Indeed, given instead $\widetilde{Y} \in \mathsf{Mod}_{\mathcal{O}[1]}$, the associated comonad arising from the adjunction (g_*Q, Ug^*) between $\mathsf{Alg}_{\mathcal{O}}$ and $\mathsf{Mod}_{\mathcal{O}[1]}$ has a natural splitting

$$g_*QUg^*(\widetilde{Y}) \simeq \bigvee_{k \geq 1} \widetilde{B}(\mathcal{O})[k] \wedge_{\mathcal{O}[1]^{\wedge k}} \widetilde{Y}^{\wedge k}$$

such that $\widetilde{B}(\mathcal{O}) \simeq B(\mathcal{O})$ via

$$\widetilde{B}(\mathcal{O}) = \tau_1\mathcal{O} \circ_{\mathcal{O}}^{\mathsf{h}} \tau_1\mathcal{O} \simeq |\,\mathrm{Bar}(\tau_1\mathcal{O}, \mathcal{O}, \tau_1\mathcal{O})| \simeq |\,\mathrm{Bar}(J, \mathcal{O}, J)| \simeq B(\mathcal{O}).$$

3.3.4   Cooperad structure on $B(\mathcal{O})$

It is known that $B(\mathcal{O})$ (resp. $\widetilde{B}(\mathcal{O})$) is a coaugmented cooperad, at least in the homotopy category of spectra (see, e.g., Ching [22] for the topological case, Lurie [60, §5] for an ∞-categorical approach, or Ginzburg-Kapranov [40] for the chain complexes case) via the natural comultiplication

$$J \circ_{\mathcal{O}}^{\mathsf{h}} J \simeq J \circ_{\mathcal{O}}^{\mathsf{h}} \mathcal{O} \circ_{\mathcal{O}}^{\mathsf{h}} J \to J \circ_{\mathcal{O}}^{\mathsf{h}} J \circ_{\mathcal{O}}^{\mathsf{h}} J \simeq (J \circ_{\mathcal{O}}^{\mathsf{h}} J) \circ_J (J \circ_{\mathcal{O}}^{\mathsf{h}} J).$$

We would like to say that the $\mathsf{Alg}_{\mathcal{O}}$ Snaith splitting allows one to immediately recognize $\partial_* \mathrm{Id}_{\mathsf{Alg}_{\mathcal{O}}}$ as the cobar construction on $B(\mathcal{O})$, however the splittings provided seem to be too weak to justify this claim (a similar problem is enocuntered in Arone-Kankaanrinta [6] for the classic Snaith splitting). As such, one benefit of our work is that we do not require any more rigid cooperad structure on $B(\mathcal{O})$ to produce our model for $\partial_* \mathrm{Id}_{\mathsf{Alg}_{\mathcal{O}}}$.

Also of note is that the $\mathsf{Alg}_{\mathcal{O}}$ Snaith splitting may be interpreted to say that any $Y \in \mathsf{Alg}_J$ (resp. $\widetilde{Y} \in \mathsf{Mod}_{\mathcal{O}[1]}$) is naturally a divided power coalgebra over $B(\mathcal{O})$ (resp. $\widetilde{B}(\mathcal{O})$), at least in the homotopy category, and that the functor $X \mapsto \mathsf{TQ}(X)$ underlies the left-adjoint to the conjectured Quillen equivalence (i.e., Koszul duality equivalence) between nilpotent $\mathcal{O}$-algebras and nilpotent divided power $B(\mathcal{O})$-coalgebras from Francis-Gaitsgory [38] (which has since been partially resolved by Ching-Harper [28]).

3.3.5 Interaction of the stabilization resolution with Taylor towers

We now provide the explicit model we employ for $\partial_* \mathrm{Id}_{\mathsf{Alg}_{\mathcal{O}}}$. Our argument is essentially to show that one can "move the ∂_* inside the holim" on the right hand side of (2.3.4) by higher stabilization and then use the $\mathsf{Alg}_{\mathcal{O}}$ Snaith splitting to recognize the resulting diagram. Let us write Id for $\mathrm{Id}_{\mathsf{Alg}_{\mathcal{O}}}$.

Proposition 3.3.6. *Let $k \geq n \geq 1$, then $P_n \mathrm{Id} \xrightarrow{\sim} \mathrm{holim}_{\Delta^{\leq k-1}} P_n((UQ)^{\bullet+1})$.*

Proof. The estimates from Proposition 2.3.9 suffice to show that the map

$$c_k \colon \mathrm{Id} \to \mathrm{holim}_{\Delta^{\leq k-1}} C(-)$$

agrees to order n on the subcategory of 0-connected objects (see [43, 1.2]) in which case $P_n(c_k)$ is an equivalence via [43, 1.6]. Further,

$$P_n(\mathrm{holim}_{\Delta^{\leq k-1}} C(-)) \simeq \mathrm{holim}_{\Delta^{\leq k-1}} P_n((UQ)^{\bullet+1})$$

as $P_n(-)$ commutes with *very finite*[11] homotopy limits by construction (cf. Section 2.3.8).

$$\square$$

Since $D_n(-)$ and $\partial_n(-)$ are built from $P_n(-)$ by very finite homotopy limits, Proposition 3.3.6 extends to an equivalence on homogeneous layers and derivatives as well. Moreover, the restriction map

$$\mathrm{holim}_{\Delta}\, P_n((UQ)^{\bullet+1}) \to \mathrm{holim}_{\Delta^{\leq k-1}}\, P_n((UQ)^{\bullet+1})$$

is an equivalence for $k \geq n \geq 1$ as the objects as a corollary to the higher stabilization estimates from Proposition 2.3.9 (resp. for D_n and ∂_n).

Let M be an $(\mathcal{O}, \mathcal{O})$-bimodule. For notational convenience, for $k \geq 1$, we set

$$M^{(k)} = \underbrace{M \circ_{\mathcal{O}} \cdots \circ_{\mathcal{O}} M}_{k}. \tag{3.3.7}$$

Note that $J^{(k)}$ is a cofibrant $(\mathcal{O}, \mathcal{O})$-bimodule with $(UQ)^{k+1}(X) = J^{(k+1)} \circ_{\mathcal{O}} (X)$. By Proposition 3.2.12, there are then equivalences

$$P_n\mathrm{Id} \xrightarrow{\sim} \mathrm{holim}_{\Delta^{\leq k-1}} \left(P_n(UQ) \Longrightarrow P_n((UQ)^2) \Lleftarrow\Rrightarrow P_n((UQ)^3) \cdots \right)$$

$$\simeq \mathrm{holim}_{\Delta^{\leq k-1}} \left(\tau_n J^{(1)} \circ_{\mathcal{O}} (-) \Longrightarrow \tau_n J^{(2)} \circ_{\mathcal{O}} (-) \Lleftarrow\Rrightarrow \tau_n J^{(3)} \circ_{\mathcal{O}} (-) \cdots \right)$$

and

$$D_n\mathrm{Id} \xrightarrow{\sim} \mathrm{holim}_{\Delta^{\leq k-1}} \left(D_n(UQ) \Longrightarrow D_n((UQ)^2) \Lleftarrow\Rrightarrow D_n((UQ)^3) \cdots \right)$$

$$\simeq \mathrm{holim}_{\Delta^{\leq k-1}} \left(i_n J^{(1)} \circ_{\mathcal{O}} (-) \Longrightarrow i_n J^{(2)} \circ_{\mathcal{O}} (-) \Lleftarrow\Rrightarrow i_n J^{(3)} \circ_{\mathcal{O}} (-) \cdots \right)$$

---

[11]Recall that a *very finite* homotopy limit is one taken over a diagram whose nerve has only finitely many nondegenerate simplices, and that such homotopy limits will commute with filtered homotopy colimits. Homotopy limits over n-cubes and punctured n-cubes are very finite

whenever $k \geq n \geq 1$.

Note there is an equivalence of restricted diagrams

$$(\tau_n J^{(\bullet+1)} \circ_{\mathcal{O}} (-))|_{\Delta^{\leq k-1}} \simeq P_n((UQ)^{\bullet+1})|_{\Delta^{\leq k-1}}$$

(resp. $(i_n J^{(\bullet+1)} \circ_{\mathcal{O}} (-))|_{\Delta^{\leq k-1}} \simeq D_n((UQ)^{\bullet+1})|_{\Delta^{\leq k-1}}$) by first replacing the coface k-cube associated to

$$\mathrm{Id} \to (UQ)^{\bullet+1}$$

by the k-cube $\mathcal{Z}_k$ (see (3.3.8) below) and then applying τ_n (resp. i_n) objectwise.

$$\{\mathcal{P}(k) \ni T \mapsto \mathcal{Z}_k(T) = (Z_1 \circ_{\mathcal{O}} \cdots \circ_{\mathcal{O}} Z_k) \circ_{\mathcal{O}} (-)\} \text{ such that } Z_i = \begin{cases} J & i \in T \\ \mathcal{O} & i \notin T \end{cases} \tag{3.3.8}$$

We then use the corresponding models for $\widetilde{\mathbb{D}}_n$ from Proposition 3.2.12 and compute the n-th derivatives via cross effects to obtain equivalences

$$\tilde{\partial}_n \mathrm{Id} \xrightarrow{\sim} \mathrm{holim}_{\Delta^{\leq k-1}} \left(\tilde{\partial}_n(UQ) \rightrightarrows \tilde{\partial}_n((UQ)^2) \Rrightarrow \tilde{\partial}_n((UQ)^3) \cdots \right) \tag{3.3.9}$$

$$\simeq \mathrm{holim}_{\Delta^{\leq k-1}} \left((\tau_1 \mathcal{O})^{(1)}[n] \rightrightarrows (\tau_1 \mathcal{O})^{(2)}[n] \Rrightarrow (\tau_1 \mathcal{O})^{(3)}[n] \cdots \right).$$

for $k \geq n \geq 1$.

Example 3.3.10. We sketch this process for $k = n = 2$. Note, there is an isomorphism of square diagrams of the form

$$\begin{array}{ccc}
\mathrm{Id} \xrightarrow{d^0} UQ & & (\mathcal{O} \circ_{\mathcal{O}} \mathcal{O}) \circ_{\mathcal{O}} (-) \xrightarrow{d^0} (\mathcal{O} \circ_{\mathcal{O}} J) \circ_{\mathcal{O}} (-) \\
\downarrow{\scriptstyle d^0} \quad \downarrow{\scriptstyle d^0} \quad \cong & & \downarrow{\scriptstyle d^0} \qquad\qquad \downarrow{\scriptstyle d^0} \\
UQ \xrightarrow{d^1} UQUQ & & (J \circ_{\mathcal{O}} \mathcal{O}) \circ_{\mathcal{O}} (-) \xrightarrow{d^1} (J \circ_{\mathcal{O}} J) \circ_{\mathcal{O}} (-).
\end{array}$$

Taking 2-homogeneous layers, we obtain an equivalence of homotopy pullback squares

$$
\begin{array}{ccc}
D_2\mathrm{Id} \xrightarrow{\ d^0\ } D_2(UQ) & & i_2(\mathcal{O} \circ_{\mathcal{O}} \mathcal{O}) \circ_{\mathcal{O}} (-) \xrightarrow{\ d^0\ } i_2(\mathcal{O} \circ_{\mathcal{O}} J) \circ_{\mathcal{O}} (-) \\
\downarrow{\scriptstyle d^0} \qquad\quad \downarrow{\scriptstyle d^0} & \simeq & \downarrow{\scriptstyle d^0} \qquad\qquad\qquad \downarrow{\scriptstyle d^0} \\
D_2(UQ) \xrightarrow{\ d^1\ } D_2(UQUQ) & & i_2(J \circ_{\mathcal{O}} \mathcal{O}) \circ_{\mathcal{O}} (-) \xrightarrow{\ d^1\ } i_2(J \circ_{\mathcal{O}} J) \circ_{\mathcal{O}} (-).
\end{array}
$$

The associated lifts $\widetilde{\mathbb{D}_2}(-)$ to functors on $\mathsf{Mod}_{\mathcal{O}[1]}$ from Proposition 3.2.12 then fit into a homotopy pullback square

$$
\begin{array}{ccc}
\widetilde{\mathbb{D}_2}\mathrm{Id} & \xrightarrow{\ d^0\ } & (\mathcal{O} \circ_{\mathcal{O}} \tau_1\mathcal{O})[2] \wedge_{\mathcal{O}[1]^{!2}} (-)^{\wedge 2} \\
\downarrow{\scriptstyle d^0} & & \downarrow{\scriptstyle d^0} \\
(\tau_1\mathcal{O} \circ_{\mathcal{O}} \mathcal{O})[2] \wedge_{\mathcal{O}[1]^{!2}} (-)^{\wedge 2} & \xrightarrow{\ d^1\ } & (\tau_1\mathcal{O} \circ_{\mathcal{O}} \tau_1\mathcal{O})[2] \wedge_{\mathcal{O}[1]^{!2}} (-)^{\wedge 2}
\end{array}
$$

which by taking cross effects cr_2 then provides an equivalence of homotopy pullback squares

$$
\begin{array}{ccccc}
\widetilde{\partial_2}\mathrm{Id} \xrightarrow{\ d^0\ } \widetilde{\partial_2}(UQ) & & \widetilde{\partial_2}\mathrm{Id} \xrightarrow{\ d^0\ } \tau_1\mathcal{O}[2] & & \partial_2\mathrm{Id} \xrightarrow{\ d^0\ } J[2] \\
\downarrow{\scriptstyle d^0} \quad\ \downarrow{\scriptstyle d^0} & \simeq & \downarrow{\scriptstyle d^0} \qquad\quad \downarrow{\scriptstyle d^0} & \simeq & \downarrow{\scriptstyle d^0} \qquad\ \downarrow{\scriptstyle d^0} \\
\widetilde{\partial_2}(UQ) \xrightarrow{\ d^1\ } \widetilde{\partial_2}(UQUQ) & & \tau_1\mathcal{O}[2] \xrightarrow{\ d^1\ } (\tau_1\mathcal{O} \circ_{\mathcal{O}} \tau_1\mathcal{O})[2] & & J[2] \xrightarrow{\ d^1\ } (J \circ_{\mathcal{O}} J)[2].
\end{array}
$$

Remark 3.3.11. It follows then that $\partial_*\mathrm{Id}$ is obtained as $\operatorname{holim}_{\mathbf{\Delta}} C(\mathcal{O}) \simeq \mathsf{Tot}\, C(\mathcal{O})$, where $C(\mathcal{O})$ is the following cosimplicial diagram (showing only coface maps)

$$
C(\mathcal{O}) = \left(J \circ_{\mathcal{O}} \mathcal{O} \rightrightarrows J \circ_{\mathcal{O}} J \circ_{\mathcal{O}} \mathcal{O} \substack{\longrightarrow\\[-0.3em]\longrightarrow\\[-0.3em]\longrightarrow} J \circ_{\mathcal{O}} J \circ_{\mathcal{O}} J \circ_{\mathcal{O}} \mathcal{O} \cdots \right) \tag{3.3.12}
$$

$$
\cong \left(J \rightrightarrows J \circ_{\mathcal{O}} J \substack{\longrightarrow\\[-0.3em]\longrightarrow\\[-0.3em]\longrightarrow} J \circ_{\mathcal{O}} J \circ_{\mathcal{O}} J \cdots \right),
$$

with coface maps as in (2.3.4), i.e., $C(\mathcal{O}) = J^{(\bullet+1)}$. In other words $C(\mathcal{O})$ provides a rigidification of the diagram $\partial_*(UQ)^{\bullet+1}$ whose terms are *a priori* defined only up to homotopy.

3.4 The Taylor tower of the identity functor in spaces

The Taylor tower of the identity is a central object of homotopy theory. From the definitions in [43], is not hard to show

$$P_1 \mathrm{Id}_{\mathsf{Top}_*}(X) \simeq D_1 \mathrm{Id}_{\mathsf{Top}_*}(X) \simeq \Omega^\infty \Sigma^\infty(X) \tag{3.4.1}$$

the stabilization of a space X. The higher Blakers-Massey theorems [42, 2.1] show that Id is 1-analytic and therefore the Taylor tower of the identity in Top_* offers an interpolation between a simply connected space $X \simeq \mathrm{holim}_n P_n \mathrm{Id}_{\mathsf{Top}_*}(X)$ and its stabilization $\Omega^\infty \Sigma^\infty X$.

The equivalences from (3.4.1) show that the first derivative of $\mathrm{Id}_{\mathsf{Top}_*}$ is just the sphere spectrum S. Johnson [52] and later Arone-Mahowald [7] give a description of the higher homogeneous layers and derivatives of $\mathrm{Id}_{\mathsf{Top}_*}$ in terms of the *partition poset complex* (Definition 3.4.5). Specifically, they show

$$D_n \mathrm{Id}_{\mathsf{Top}_*}(X) \simeq \Omega^\infty \left(\mathrm{Map}\left(\mathrm{Par}(n), \Sigma^\infty X^{\wedge n}\right)_{h\Sigma_n} \right) \tag{3.4.2}$$

and similarly

$$\partial_n \mathrm{Id}_{\mathsf{Top}_*} \simeq \mathrm{Map}(\mathrm{Par}(n), S) \tag{3.4.3}$$

for all $n \geq 1$. In particular, the n-th derivative of $\mathrm{Id}_{\mathsf{Top}_*}$ is just the Spanier-Whitehead dual to $\mathrm{Par}(n)$.

3.4.4 The partition poset complex

For $n \geq 0$ we denote by $\mathbf{n}$ the set $\{1, \ldots, n\}$, note that $\mathbf{0} = \emptyset$. A *partition* λ of $\mathbf{n}$ is a decomposition $\mathbf{n} = \coprod_{i \in I} T_i$ into nonempty subsets (here I is required to be a nonempty set). Given partition $\lambda = \{T_i\}_{i \in I}$ and $\lambda' = \{T_j'\}_{j \in J}$ of $\mathbf{n}$ we say that $\lambda \leq \lambda'$ if there is a surjection

$f\colon J \to I$ such that $T_i = \coprod_{j \in f^{-1}(i)} T'_j$ for all $i \in I$.

Note that the set of partitions of $\mathbf{n}$ has a minimal element min consisting of only the trivial partition $\{1, \ldots, n\}$, and a maximal element max consisting of the partition of $\mathbf{n}$ into singletons, i.e. $\{\{1\}, \ldots, \{n\}\}$. The set of partitions of $\mathbf{n}$ then forms a poset with respect to $\leq$, and so may be interpreted as a category. The partition poset complex as defined below is (a quotient of) the nerve of this category.

Definition 3.4.5. Define the *n-th partition poset complex* $\mathsf{Par}(n)$ to be the geometric realization of the pointed simplicial set $P(n)$ defined as follows. The k-simplices of $P(n)$ are given by sequences
$$\lambda_0 \leq \lambda_1 \leq \cdots \leq \lambda_{k-1} \leq \lambda_k$$
of partitions of $\mathbf{n}$ such that any chain that does not satisfy $\lambda_0 = \mathsf{min}$ and $\lambda_k = \mathsf{max}$ is identified with the basepoint.

Face maps $d_i\colon P(n)_k \to P(n)_{k-1}$ are given by removing the i-th entry λ_i and degeneracy maps $s_j\colon P(n)_k \to P(n)_{k+1}$ are given by repeating the j-th entry λ_j. Note, that the image of d_0 (resp. d_k) is only the basepoint if $\lambda_1 \neq \mathsf{min}$ (resp. $\lambda_{k-1} \neq \mathsf{max}$).

More generally, for a finite set T we define $\mathsf{Par}(T)$ analogously, e.g. by setting $|T| = n$ and specifying a bijection $T \cong \mathbf{n}$.

Remark 3.4.6. Note that $\mathsf{Par}(n)$ inherits a natural action of Σ_n by permuting the elements of $\mathbf{n}$. A useful observation is that non-basepoint elements $\alpha \in P(n)_k$ are in bijective correspondence with isomorphism classes of planar, rooted trees with n labelled leaves and k levels, up to planar isomorphism.

Example 3.4.7. It is possible to calculate some low dimensional examples of partition poset complexes. For instance, $\mathsf{Par}(0) = *$, $\mathsf{Par}(1) \cong S^0$ and $\mathsf{Par}(2) \cong S^1$ with trivial Σ_2 action. Similarly, $\mathsf{Par}(3)$ may be identified with the 2-sphere with a disc glued-in at the equator, Σ_3 acts on $\mathsf{Par}(3)$ by permuting the three 2-discs (top hemisphere, bottom hemisphere and equator).

Moreover, it is known that there is a (nonequivaraint) equivalence (see [52], [7])

$$\mathsf{Par}(n) \simeq \bigvee_{i=1}^{(n-1)!} S^{n-1}.$$

We briefly describe one route of arriving at the model (3.4.3), using the approach of Arone-Kankaanrinta [6] to analyze $\mathsf{Id}_{\mathsf{Top}_*}$ by the Snaith splitting.

3.4.8 Analysis of the Taylor tower of the identity by higher stabilization

Associated to the stabilization adjunction $(\Sigma^\infty, \Omega^\infty)$ between Top_* and Spt, for any space X, there is a coaugmented cosimplicial diagram $X \to C(\Sigma^\infty X)$. Here, $C(\Sigma^\infty X)$ is the cobar resolution

$$C(\Sigma^\infty X) := \mathrm{Cobar}(\Omega^\infty, \Sigma^\infty \Omega^\infty, \Sigma^\infty X)$$

and the coaugmented is provided by the unit map $X \to \Omega^\infty \Sigma^\infty X$. $C(\Sigma^\infty X)$ is functorial in X and provides a cosimplicial functor

$$C(\Sigma^\infty -) = \left(\Omega^\infty \Sigma^\infty \rightrightarrows (\Omega^\infty \Sigma^\infty)^2 \substack{\longrightarrow\\[-0.6em]\longrightarrow\\[-0.6em]\longrightarrow} (\Omega^\infty \Sigma^\infty)^3 \cdots \right) \tag{3.4.9}$$

whose coface maps are induced by inserting the unit map $\mathsf{Id}_{\mathsf{Top}_*} \to Q := \Omega^\infty \Sigma^\infty$ and codegeneracy maps are induced by the counit map $\Sigma^\infty \Omega^\infty =: \mathsf{K} \to \mathsf{Id}_{\mathsf{Spt}}$.

Blomquist-Harper [15] utilize the cubical higher Blakers-Massey theorems of [42] to recover a classical result of Bousfield-Kan [19] (see also [21]) that simply connected spaces are equivalent to their completion with respect to stabilization. Specifically, if $X \in \mathsf{Top}_*$ is 1-connected, then

$$X \xrightarrow{\sim} \mathrm{holim}_\Delta\, C(X) \simeq X^\wedge_{\Omega^\infty \Sigma^\infty}.$$

The key to their proof is strong connectivity estimates of the following form.

Proposition 3.4.10. *The comparison map $X \to \mathrm{holim}_{\Delta^{\leq k-1}} C(\Sigma^\infty X)$ is $(c(k+1)+1)$-connected for $X \in \mathsf{Top}_*$ c-connected.*

We make use of the above connectivity estimates to show that $P_n \mathrm{Id}$ may be recovered as the totalization of $P_n(Q^{\bullet+1})$ (see also [2, §16]).

Corollary 3.4.11. *Let $k \geq n \geq 1$, then $P_n \mathrm{Id}_{\mathsf{Top}_*} \xrightarrow{\sim} \mathrm{holim}_{\Delta^{\leq k-1}} P_n(Q^{\bullet+1})$.*

Proof. Using the estimates from Proposition 3.4.10, this follows from the same argument as in Proposition 3.3.6. $\qquad\square$

The above corollary readily extends to equivalences on D_n and ∂_n as well. The upshot for us is that $\partial_n(Q^{k+1})$ is readily computable via the Snaith splitting, as follows.

3.4.12 The Snaith splitting

Let $\underline{S}$ denote the symmteric sequence in Spt such that $\underline{S}[n] = S$ with trivial Σ_n action. The *Snaith splitting* (see e.g., [69], [30]) provides equivalences

$$\Sigma^\infty \Omega^\infty \Sigma^\infty X \simeq \bigvee_{k \geq 1} \Sigma^\infty X^{\wedge k}_{h\Sigma_k} \simeq \bigvee_{k \geq 1} S \wedge_{\Sigma_k} (\Sigma^\infty X)^{\wedge k} \tag{3.4.13}$$

Said differently, the Taylor tower for $\mathsf{K} = \Omega^\infty \Sigma^\infty$ splits as a coproduct of its homogeneous layers when evaluated on a suspension spectrum and that $\partial_* \mathsf{K} \simeq \underline{S}$.

A result of Arone-Kankaanrinta [6] uses the above splittings to recover the model for n-th homogeneous layers and n-th derivatives of the identity in spaces in (3.4.2) and (3.4.3), respectively. The crux of their argument is that iterating the Snaith splitting provides equivalences

$$\partial_n(Q^{k+1}) \simeq \partial_n(\mathsf{K}^k) \simeq \underline{S}^{\circ k}[n].$$

Here, $\underline{S}^{\circ k}$ denotes the k-fold composition of the symmetric sequence $\underline{S}$.

Remark 3.4.14. A key observation is that $\underline{S}^{\circ k}[n]$ is just a wedge of copies of the sphere spectrum S indexed by the k-simplices of $P(n)$. This symmetric sequence $\underline{S}$ admits both an operad and *cooperad* structure, the derivatives of the identity may be further recognized as the totalization of the cobar complex $C(\underline{S})$ with respect to this cooperad structure. We defer a discussion of cooperads and their coalgebras to Appendix B. In Section 6.2 we show that $C(\underline{S})$ inherits the structure of a $\overset{\circ}{\square}$-monoid (see 4.2.3), which induces the desired operad structure upon passing to totalization.

Chapter 4

The box product and box monoids

4.1 The box-product of cosimplicial objects

The aim of this section is to introduce the box product $\square$ for cosimplicial objects in a monoidal category $(\mathsf{C}, \otimes, \mathbf{1})$ as first introduced by Batanin [10]. For nice categories C (e.g., C closed, symmetric monoidal), the box product endows $\mathsf{C}^{\mathbf{\Delta}}$ with a monoidal structure, and cosimplicial objects which admits a monoidal pairing with respect to $\square$ inherit an A_∞-monoidal pairing on their totalizations (see, e.g., McClure-Smith [65, 3.1]).

Our main use of the box product will be to produce a homotopy-coherent (i.e., A_∞-) composition on the derivatives of the identity in $\mathcal{O}$-algebras, modeled as $\mathsf{Tot}\, C(\mathcal{O})$, by demonstrating a natural pairing $C(\mathcal{O})\square C(\mathcal{O}) \to C(\mathcal{O})$ (Example 4.1.4).

Definition 4.1.1. Let $(\mathsf{C}, \otimes, \mathbf{1})$ be a monoidal category and $X, Y \in \mathsf{C}^{\mathbf{\Delta}}$. Define their *box product* $X\square Y \in \mathsf{C}^{\mathbf{\Delta}}$ at level n by

$$(X\square Y)^n := \mathrm{colim}\left(\coprod_{p+q=n} X^p \otimes Y^q \rightrightarrows \coprod_{r+s=n-1} X^r \otimes Y^s \right)$$

where the maps are induced by $\mathrm{id} \otimes d^0$ and $d^{r+1} \otimes \mathrm{id}$. The object $X\square Y$ inherits cosimplicial structure via coface maps $d^i \colon (X\square Y)^n \to (X\square Y)^{n+1}$ induced by

$$\begin{cases} X^p \otimes Y^q \xrightarrow{d^i \otimes \mathrm{id}} X^{p+1} \otimes Y^q & i \leq p \\ X^p \otimes Y^q \xrightarrow{\mathrm{id} \otimes d^{i-p}} X^p \otimes Y^{q+1} & i > p \end{cases}$$

and codegeneracy maps $s^j \colon (X\square Y)^n \to (X\square Y)^{n-1}$ induced by

$$\begin{cases} X^p \otimes Y^q \xrightarrow{s^j \otimes \mathrm{id}} X^{p-1} \otimes Y^q & j < p \\ X^p \otimes Y^q \xrightarrow{\mathrm{id} \otimes s^{j-p}} X^p \otimes Y^{q-1} & j \geq p \end{cases}$$

see also Ching-Harper [28, §4].

Remark 4.1.2. Note, $(X\square Y)^0 \cong X^0 \otimes Y^0$, $(X\square Y)^1$ and $(X\square Y)^2$ may be computed as the colimits of

$$\begin{array}{ccc}
X^0 \otimes Y^1 & & X^0 \otimes Y^2 \\
\uparrow{\scriptstyle \mathrm{id} \otimes d^0} & & \uparrow{\scriptstyle \mathrm{id} \otimes d^0} \\
X^0 \otimes Y^0 \xrightarrow{d^1 \otimes \mathrm{id}} X^1 \otimes Y^0 & \text{and} & X^0 \otimes Y^1 \xrightarrow{d^1 \otimes \mathrm{id}} X^1 \otimes Y^1 \\
& & \uparrow{\scriptstyle \mathrm{id} \otimes d^0} \\
& & X^1 \otimes Y^0 \xrightarrow{d^2 \otimes \mathrm{id}} X^2 \otimes Y^0
\end{array}$$

respectively, and in general $(X\square Y)^n$ may be computed as the colimit of the staircase diagram

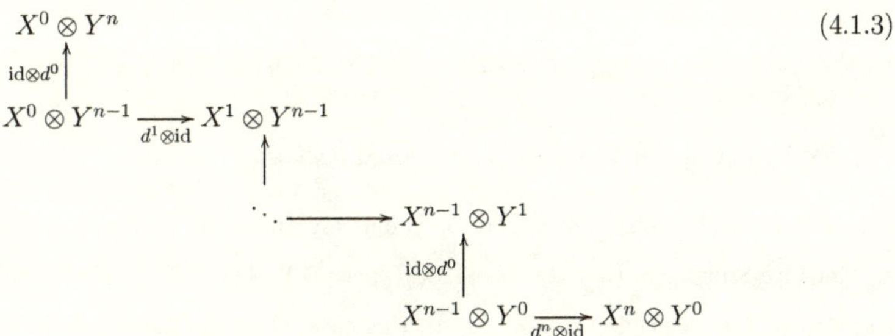

$$\tag{4.1.3}$$

In particular, if $(\mathsf{C}, \otimes, \mathbf{1})$ is closed, symmetric monoidal then $\square$ defines a monoidal category $(\mathsf{C}^{\Delta}, \square, \underline{\mathbf{1}})$, here $\underline{\mathbf{1}}$ is the constant cosimplicial object on the unit $\mathbf{1} \in \mathsf{C}$ (see, e.g., Batanin [10]).

Example 4.1.4. Recall the cosimplicial symmetric sequence $C(\mathcal{O}) = J^{(\bullet+1)}$ from (3.3.12). We observe that $C(\mathcal{O})$ admits a pairing $C(\mathcal{O})\,\overset{\circ}{\square}\,C(\mathcal{O}) \overset{m}{\to} C(\mathcal{O})$, where $\overset{\circ}{\square}$ denotes the box product in $\mathsf{SymSeq}^{\Delta}_{\mathsf{Spt}}$, induced as follows. Let c denote the operad composition map $c\colon J \circ J \to J$. Then,

$$(C(\mathcal{O})\,\overset{\circ}{\square}\,C(\mathcal{O}))^0 \cong J \circ J \overset{c}{\to} J = C(\mathcal{O})^0$$

For level 1 we observe that there are maps

$$m_{0,1}\colon J \circ J \circ_{\mathcal{O}} J \to J \circ_{\mathcal{O}} J \quad \text{and} \quad m_{1,0}\colon J \circ_{\mathcal{O}} J \circ J \to J \circ_{\mathcal{O}} J$$

induced by $J \circ J \to J$ which induces m via the following commuting square

$$\tag{4.1.5}$$

More generally, there are maps of the form

$$m_{p,q}\colon J^{(p)} \circ J^{(q)} \to J^{(p+q)} \ \text{ for } p+q=n,\ p,q \geq 0$$

induced by c, which induces the pairing m at level n.

Remark 4.1.6. The above construction is entirely analogous to the following example found in McClure-Smith [65] that the based loop space ΩX of $X \in \mathsf{Top}_*$ admits an A_∞ composition induced by an underlying $\square$-pairing. In this case, ΩX is modeled as the totalization of the cobar complex $c(X)$ built with respect to the natural comultiplication (with coaugmentation) given by the diagonal $\delta\colon X \to X \times X$.

It follows that $c(X)^p \cong X^{\times p}$ and the pairing $c(X)\square c(X) \to c(X)$ is induced by the natural isomorphisms $X^{\times p} \times X^{\times q} \cong X^{\times p+q}$. Further, McClure-Smith show that $\mathsf{Tot}\, c(X)$ is an algebra over the (nonsymmetric) coendmorphism operad on $\Delta^\bullet$, i.e.,

$$\mathbb{A}[n] = \mathrm{Map}^{\mathsf{Top}}_{\Delta^\bullet_{\mathrm{res}}} \left(\Delta^\bullet, (\Delta^\bullet)^{\square n} \right)$$

which satisfies $\mathbb{A}[0] = *$ and $\mathbb{A}[n] \xrightarrow{\sim} *$ for $n \geq 1$ (in fact Δ^n and $(\Delta^\bullet \square \Delta^\bullet)^n$ are *homeomorphic*), and that with respect to this structure $\mathsf{Tot}\, c(X) \simeq \Omega X$ as A_∞-monoids.

4.2 The box product in SymSeq^Δ

Our aim now is to build a framework in which we can work with the structure captured by Example 4.1.4, e.g., by considering the box-product in the category of cosimplicial objects in $(\mathsf{SymSeq}_\mathsf{C}, \circ, I)$ of symmetric sequences for $(\mathsf{C}, \otimes, \mathbf{1})$ some closed symmetric monoidal category.

The main difficulty is that the composition product of symmetric sequences does not always commute with colimits taken in the right hand entry. That is, for $B\colon \mathcal{I} \to \mathsf{SymSeq}_\mathsf{C}$

a small diagram and $A \in \mathsf{SymSeq}_\mathsf{C}$, the universal map

$$\mathrm{colim}_{i \in \mathcal{I}}(A \circ B_i) \to A \circ (\mathrm{colim}_{i \in \mathcal{I}} B_i) \tag{4.2.1}$$

is not an isomorphism in general. Thus the box-product fails to be strictly monoidal in this setting.

Let us write $\mathsf{SymSeq} = \mathsf{SymSeq}_\mathsf{C}$ and $\mathring{\square}$ for the box-product in SymSeq^Δ (in words we refer to $\mathring{\square}$ as the *box-circle product*). Let $\mathcal{X}, \mathcal{Y}, \mathcal{Z}, \ldots$ be cosimplicial symmetric sequences. We will systematically interpret expressions of the form $\mathcal{X} \mathring{\square} \mathcal{Y} \mathring{\square} \mathcal{Z}$ to be expanded *from the left*, i.e.,

$$\mathcal{X} \mathring{\square} \mathcal{Y} \mathring{\square} \mathcal{Z} := (\mathcal{X} \mathring{\square} \mathcal{Y}) \mathring{\square} \mathcal{Z}, \quad \mathcal{X} \mathring{\square} \mathcal{Y} \mathring{\square} \mathcal{Z} \mathring{\square} \mathcal{W} := ((\mathcal{X} \mathring{\square} \mathcal{Y}) \mathring{\square} \mathcal{Z}) \mathring{\square} \mathcal{W}, \ \ldots$$

and note that via the universal map in (4.2.1) there is always a canonical comparison map θ of the form

$$\theta \colon \mathcal{X} \mathring{\square} \mathcal{Y} \mathring{\square} \mathcal{Z} = (\mathcal{X} \mathring{\square} \mathcal{Y}) \mathring{\square} \mathcal{Z} \to \mathcal{X} \mathring{\square} (\mathcal{Y} \mathring{\square} \mathcal{Z}) \tag{4.2.2}$$

which likely fails to be invertible. However, θ is sufficient to provide a suitable description of monoids with respect to $\mathring{\square}$, i.e., Definition 4.2.3, below. First, we note that the unit $I \in \mathsf{SymSeq}_\mathsf{C}$ induces a unit $\underline{I} \in \mathsf{SymSeq}_\mathsf{C}$ as the constant cosimplicial object on I in that there are isomorphisms

$$\mathcal{X} \mathring{\square} \underline{I} \cong \mathcal{X} \cong \underline{I} \mathring{\square} \mathcal{X}.$$

For instance, the right isomorphism is obtained by noting that for any p, q the map $d^{p+1} \circ \mathrm{id}$ in the following

$$
\begin{array}{c}
\underline{I}^p \circ \mathcal{X}^{q+1} \\
\uparrow{\scriptstyle \mathrm{id} \circ d^0} \\
\underline{I}^p \circ \mathcal{X}^q \xrightarrow{d^{p+1} \circ \mathrm{id}} \underline{I}^{p+1} \circ \mathcal{X}^q
\end{array}
$$

is just the identity (and hence has an inverse). Therefore, the inclusion of the vertex $\underline{I}^0 \circ \mathcal{X}^n$

into the diagram defining $(\mathcal{X}\,\mathring{\square}\,\mathcal{X})^n$ is right cofinal (i.e., induces an isomorphism on colimits).

Definition 4.2.3. By $\mathring{\square}$-monoid in SymSeq^{Δ}, we mean a cosimplicial symmetric sequence $\mathcal{X}$ together with maps $m\colon \mathcal{X}\,\mathring{\square}\,\mathcal{X} \to \mathcal{X}$ and $u\colon \underline{I} \to \mathcal{X}$ so that the following associativity (4.2.4) and unitality (4.2.5) diagrams commute

$$
\begin{array}{ccccc}
\mathcal{X}\,\mathring{\square}\,\mathcal{X}\,\mathring{\square}\,\mathcal{X} & \xrightarrow{\ \theta\ } & \mathcal{X}\,\mathring{\square}\,(\mathcal{X}\,\mathring{\square}\,\mathcal{X}) & \xrightarrow{\mathrm{id}\mathring{\square}m} & \mathcal{X}\,\mathring{\square}\,\mathcal{X} \\
\Big\| = & & & & \Big\downarrow m \\
(\mathcal{X}\,\mathring{\square}\,\mathcal{X})\,\mathring{\square}\,\mathcal{X} & \xrightarrow{m\mathring{\square}\mathrm{id}} & \mathcal{X}\,\mathring{\square}\,\mathcal{X} & \xrightarrow{\ m\ } & \mathcal{X}
\end{array}
\tag{4.2.4}
$$

and

$$
\begin{array}{ccc}
\mathcal{X}\,\mathring{\square}\,\underline{I} \xrightarrow{\ \mathrm{id}\mathring{\square}u\ } & \mathcal{X}\,\mathring{\square}\,\mathcal{X} & \xleftarrow{\ u\mathring{\square}\mathrm{id}\ } \underline{I}\,\mathring{\square}\,\mathcal{X} \\
{\scriptstyle\cong}\searrow \quad & \Big\downarrow m & \quad \swarrow{\scriptstyle\cong} \\
& \mathcal{X} &
\end{array}
\tag{4.2.5}
$$

Remark 4.2.6. We remark that in the language of Ching [25] (see also Day-Street [31]), $\mathring{\square}$ admits a *normal oplax monoidal structure* by defining

$$
\mathcal{X}_1\,\mathring{\square}\cdots\mathring{\square}\,\mathcal{X}_k := (\cdots((\mathcal{X}_1\,\mathring{\square}\,\mathcal{X}_2)\,\mathring{\square}\,\mathcal{X}_3)\cdots)\,\mathring{\square}\,\mathcal{X}_k
$$

and obtaining grouping maps from the universal map in (4.2.1). Our notion of $\mathring{\square}$-monoids are *normal oplax monoids* with respect to such structure by appealing to Ching [25, 3.4], noting in particular that four-fold and higher associativity diagrams are known to commute given the commutativity of (4.2.4).

Proposition 4.2.7. *The cosimplicial symmetric sequence $C(\mathcal{O})$ (see (3.3.12)) admits a natural $\mathring{\square}$-monoid structure, i.e., there are maps $m\colon C(\mathcal{O})\,\mathring{\square}\,C(\mathcal{O}) \to C(\mathcal{O})$ and $u\colon \underline{I} \to C(\mathcal{O})$ which satisfy associativity and unitality.*

Proof. The map m is that constructed in Example 4.1.4. The unit $I \to J$ provides a coaugmentation $I \to C(\mathcal{O})$ which in turn induces a map $u\colon \underline{I} \to C(\mathcal{O})$.

Associativity (4.2.4) follows from a routine calculation, observing that

$$d^0 \colon (C(\mathcal{O}) \mathring{\square} C(\mathcal{O}))^q \to (C(\mathcal{O}) \mathring{\square} C(\mathcal{O}))^{q+1}$$

is induced by $d^0 \circ \mathrm{id} \colon C(\mathcal{O})^r \circ C(\mathcal{O})^s \to C(\mathcal{O})^{r+1} \circ C(\mathcal{O})^s$ for $r + s = q$. Similarly, the right-hand triangle from the unitality diagram (4.2.5) is granted by the following commuting diagrams

$$
\begin{array}{ccc}
\underline{I}^p \circ C(\mathcal{O})^q & \xrightarrow{\ u^p \circ\, \mathrm{id}\ } & C(\mathcal{O})^p \circ C(\mathcal{O})^q \\
{\scriptstyle \cong}\big\downarrow & & \big\downarrow{\scriptstyle m_{p,q}} \\
C(\mathcal{O})^q & \xrightarrow{\ d^0 \cdots d^0\ } & C(\mathcal{O})^{p+q}
\end{array}
$$

for all p, q. A similar argument provides the commutativity of the other side of the unitality diagram. $\qquad\square$

Theorem 1.1.1(a) is then obtained as a corollary to the following proposition, the proof of which is deferred to Section 6.1.1. As such, the aim of the following sections is to set up a precise framework to describe what is meant by A_∞-operad.

Proposition 4.2.8. *If $\mathcal{X}$ is a $\mathring{\square}$-monoid in $\mathsf{SymSeq}^{\triangle}_{\mathsf{Spt}}$, then $\mathsf{Tot}\,\mathcal{X}$ is an A_∞-monoid with respect to the composition product (i.e., A_∞-operad).*

Chapter 5

N-colored operads with levels

In this section we develop our theory of $\mathbf{N} = \{0, 1, 2, \dots\}$-colored operads with levels, which we refer to as $\mathbf{N}_{\mathsf{lev}}$-operads. The motivating principle behind our constructions is to provide a framework to fatten-up the usual notion of operads and their algebras. For this section $(\mathsf{C}, \otimes, \mathbf{1})$ will denote a given cocomplete closed, symmetric monoidal category with initial object ϕ. We first recall the classical theory of colored operads.

5.1 Colored operads

Colored operads (sometimes also referred to as *multicategories*) offer a generalization of operads to encode more nuanced algebraic operations on their algebras. We give an overview of their pertinent details below and refer the reader to Leinster [59] or Elmendorf-Mandell [35] for more information. As before, we will only need to consider colored operads in the category of spectra.

Definition 5.1.1. Let C be a nonempty set, i.e., a set of *colors*. A *C-colored operad* $\mathcal{M}$ in C consists of

- Objects $\mathcal{M}(c_1, \dots, c_n; d) \in \mathsf{C}$ for all $(c_1, \dots, c_n; d) \in C^{\times n} \times C$ and $n \geq 0$

- A unit map $1 \to \mathcal{M}(c; c)$ for all $c \in C$

- Composition maps of the form

$$\mathcal{M}(c_1, \ldots, c_n; d) \otimes \mathcal{M}(p_{1,1}, \ldots, p_{1,k_1}; c_1) \otimes \cdots \otimes \mathcal{M}(p_{n,1}, \ldots, p_{n,k_n}; c_n) \qquad (5.1.2)$$
$$\to \mathcal{M}(p_{1,1}, \ldots, p_{n,k_n}; d)$$

subject equivariance, associativity and unitality conditions (see, e.g., [35, 2.1]).

An *algebra* over $\mathcal{M}$ is a C-colored object, i.e., $X = \{X_c\}_{c \in C}$ such that $X_c \in \mathsf{C}$ for all $c \in C$, together with maps for each tuple $(c_1, \ldots, c_n; d) \in C^{\times n} \times C$ of the form

$$\mathcal{M}(c_1, \ldots, c_n; d) \otimes X_{c_1} \otimes \cdots \otimes X_{c_n} \to X_d$$

the collection of which is required to satisfy equivariance, associativity and unitality conditions.

Berger-Moerdijk provide a list of examples in [13, §1.5]; of note is that for $C = \{*\}$, a one-colored operad is just an operad in the classical sense. The following constructions are also motivated by White-Yau [70] wherein a composition product for C-colored operads is provided.

5.2 $\mathsf{N}_{\mathsf{lev}}$-objects

The purpose of this section is to introduce the notion of a nonsymmetric, $\mathbf{N}$-colored sequence *with levels* in C. We will refer to these as $\mathsf{N}_{\mathsf{lev}}$-*objects*. In our framework, $\mathsf{N}_{\mathsf{lev}}$-objects will play a role analogous to symmetric sequences for classical (one-color) operads, though we note that we do not yet impose any symmetric group actions on our $\mathsf{N}_{\mathsf{lev}}$-objects. Let $\mathbf{s}$ denote the set $\{1, \ldots, s\}$ (note that $\mathbf{0} = \emptyset$).

Definition 5.2.1. For $k \geq 0$, let $\mathbf{N}^{\hat{o}k}$ denote the set of tuples of orbits

$$\mathbf{N}^{\hat{o}k} := \left\{ \left(n^1, (n_1^2, \cdots, n_{n^1}^2)_{\Sigma_{n^1}}, \cdots, (n_1^k, \cdots, n_{n^{k-1}}^k)_{\Sigma_{n^{k-1}}} \right) : n_i^j \geq 0 \; \forall i, j \right\}$$

where n^j is inductively defined as $\sum_{i=1}^{n^{j-1}} n_i^j$ and we set $n^0 := 1$. We then treat $\mathbf{N}^{\hat{o}k}$ as a category with only identity morphisms.

Note that the superscripts in Definition 5.2.1 are used for indexing and are not powers, we will adhere to this convention throughout the document. Elements $\underline{p} \in \mathbf{N}^{\hat{o}k}$ will be referred to as *profiles*, we will often suppress the orbit subscript and write $(n_1, \ldots, n_s)$ for the orbit $(n_1, \ldots, n_s)_{\Sigma_s}$.

Definition 5.2.2. Given $\underline{p} = (n^1, \ldots, (n_i^k)_{i \in \mathbf{n}^{k-1}}) \in \mathbf{N}^{\hat{o}k}$, we define the *weight* of $\underline{p}$ to be the integer $n^k = \sum_{i \in \mathbf{n}^{k-1}} n_i^k$. For $t \in \mathbf{N}$, we write $\mathbf{N}^{\hat{o}k}[t]$ for the set of profiles $\underline{p} \in \mathbf{N}^{\hat{o}k}$ of weight t.

Example 5.2.3. Computing small examples we see

$$\mathbf{N}^{\hat{o}0} = \{\emptyset\}, \qquad \mathbf{N}^{\hat{o}2} \cong \{(n, (k_1, \ldots, k_n)) : n, k_i \geq 0\},$$

$$\mathbf{N}^{\hat{o}1} \cong \mathbf{N}, \qquad \mathbf{N}^{\hat{o}3} \cong \{(n, (k_1, \ldots, k_n), (t_1, \ldots, t_k)) : k = k_1 + \cdots + k_n, n, k_i, t_j \geq 0\}.$$

Remark 5.2.4. Note that profiles in $\mathbf{N}^{\hat{o}\ell}$ are in bijective correspondence to indexing factors of ℓ-fold iterates of $\hat{o}$ from (2.1.7), therefore objects indexed on $\mathbf{N}^{\hat{o}\ell}$ naturally arise when evaluating ℓ-fold iterates of the composition product of symmetric sequences (Definition 2.1.6) from the left.

Given $\underline{p} = (n^1, (n_i^2)_{i \in \mathbf{n}^1} \ldots, (n_i^\ell)_{i \in \mathbf{n}^{\ell-1}}) \in \mathbf{N}^{\hat{o}\ell}[t]$, the term

$$(X_1 \circ \cdots \circ X_\ell)[\underline{p}]$$

is the collection of factors in $(X_1 \circ \cdots \circ X_\ell)[t]$ corresponding to the indexing tuples

$$(n_1^j, \ldots, n_{n^1}^j)_{\Sigma_{n^{j-1}}} \in \mathsf{Sum}_{n^{j-1}}^{n^j},$$

for $j = 1, \ldots, \ell$.

Definition 5.2.5. Given profiles $\underline{p}, \underline{q} \in \mathbf{N}^{\hat{o}k}$ we define their *amalgamation* $\underline{p} \amalg \underline{q}$ to be the orbit of the levelwise disjoint union of the two profiles. In other words, given

$$\underline{p} = (n^1, (n_i^2)_{i \in \mathbf{n^1}}, (n_i^3)_{i \in \mathbf{n^2}}, \ldots, (n_i^k)_{i \in \mathbf{n^{k-1}}}),$$
$$\underline{q} = (m^1, (m_j^2)_{j \in \mathbf{m^1}}, (m_j^3)_{j \in \mathbf{m^2}}, \ldots, (m_j^k)_{j \in \mathbf{m^{k-1}}}),$$

then $\underline{p} \amalg \underline{q}$ is given by

$$\underline{p} \amalg \underline{q} := \Big((n^1, m^1), ((n_i^2)_{i \in \mathbf{n^1}} \amalg (m_j^2)_{j \in \mathbf{m^1}}), \ldots, ((n_i^k)_{i \in \mathbf{n^{k-1}}} \amalg (m_j^k)_{j \in \mathbf{m^{k-1}}}) \Big).$$

Remark 5.2.6. Note that $\underline{p} \amalg \underline{q}$ is *not* an element of any $\mathbf{N}^{\hat{o}k}$ as its first entry is not a singleton. However, if $\underline{p}_i \in \mathbf{N}^{\hat{o}k}[t_i]$ for $i = 1, \ldots, n$ then

$$\big(n, \underline{p_1} \amalg \cdots \amalg \underline{p_n}\big) \in \mathbf{N}^{\hat{o}k+1}[t_1 + \cdots + t_n].$$

For instance, if $\underline{p} = (2, (2, 3))$ and $\underline{q} = (3, (2, 3, 4))$ then

$$\big(2, \underline{p} \amalg \underline{q}\big) = \big(2, (2, 3)_{\Sigma_2}, (2, 3, 2, 3, 4)_{\Sigma_5}\big) \in \mathbf{N}^{\hat{o}3}[14].$$

Definition 5.2.7. An $\mathbf{N}_{\mathsf{lev}}$-*object* $\mathcal{P}$ in a symmetric monoidal category C is a functor

$$\mathcal{P} \colon \coprod_{\ell \geq 0} \mathbf{N}^{\hat{o}\ell} \times \mathbf{N} \to \mathsf{C}.$$

Equivalently, $\mathcal{P} = (\mathcal{P}_k)_{k \geq 0}$ such that $\mathcal{P}_k$ is a functor $\mathbf{N}^{\hat{\circ}k} \times \mathbf{N} \to \mathsf{C}$. We also refer to $\mathbf{N}_{\mathsf{lev}}$-objects as $\mathbf{N}$-*colored objects with levels*. We further say an $\mathbf{N}_{\mathsf{lev}}$-object $\mathcal{P}$ is *reduced* if

- For $\ell \geq 1$, $\mathcal{P}_\ell(\underline{p}; t) = \boldsymbol{\phi}$ if $\underline{p} \notin \mathbf{N}^{\hat{\circ}\ell}[t]$

- $\mathcal{P}_0(\emptyset; 1) = \mathbf{1}$

- $\mathcal{P}_0(\emptyset; n) = \boldsymbol{\phi}$ for $n \neq 1$.

Recall that $\boldsymbol{\phi}$ denotes the initial object of C.

Note if $\mathcal{P}$ is reduced then $\mathcal{P}$ is determined by a functor $\coprod_{\ell \geq 0} \mathbf{N}^{\hat{\circ}\ell} \to \mathsf{C}$. We will mostly be concerned with reduced $\mathbf{N}_{\mathsf{lev}}$-objects, but benefit from this more general definition when we discuss algebras in Section 5.3.6.

5.2.8 A composition product for $\mathbf{N}_{\mathsf{lev}}$-objects

The aim of this section is develop a monoidal composition product for $\mathbf{N}_{\mathsf{lev}}$-objects so that we may encode $\mathbf{N}_{\mathsf{lev}}$-operads as monoids.

Definition 5.2.9. Let $\underline{p} = (n^1, (n_i^2)_{i \in \mathbf{n}^1}, \ldots, (n_i^k)_{i \in \mathbf{n}^{k-1}}) \in \mathbf{N}^{\hat{\circ}k}$ and let $\ell_1, \ldots, \ell_k \geq 0$ be given. Let $\underline{Q}$ denote a collection of unordered sequences of profiles $(\underline{q}_1^j, \cdots, \underline{q}_{n^j-1}^j)$ for $j = 1, \ldots, k$ such that $\underline{q}_i^j \in \mathbf{N}^{\hat{\circ}\ell_j}[n_i^j]$.

We define the *composite* of $\underline{p}$ and $\underline{Q}$ to be the profile $\underline{p} \circ \underline{Q} \in \mathbf{N}^{\hat{\circ}(\ell_1 + \cdots + \ell_k)}$ given as follows

$$\underline{p} \circ \underline{Q} := \left(\underline{q}^1, \underline{q}_1^2 \amalg \cdots \amalg \underline{q}_{n^1}^2, \cdots, \underline{q}_1^k \amalg \cdots \amalg \underline{q}_{n^{k-1}}^k \right).$$

Let

$$\mathbf{N}^{\hat{\circ}k} \ltimes \left(\coprod_{\ell_1, \cdots, \ell_k \geq 0} \mathbf{N}^{\hat{\circ}\ell_1} \times \cdots \times \mathbf{N}^{\hat{\circ}\ell_k} \right)$$

be the collection of all pairs $(\underline{p}, \underline{Q})$ such that

$$\underline{p} = (n^1, (n_i^2)_{i \in \mathbf{n^1}}, \ldots, (n_i^k)_{i \in \mathbf{n^{k-1}}}), \qquad \underline{Q} = (\underline{q}^1, \cdots, (\underline{q}_j^k)_{j \in \mathbf{n^{k-1}}})$$

so that the composite $\underline{p} \circ \underline{Q}$ is defined (i.e., $\underline{q}_i^j \in \mathbf{N}^{\hat{\mathrm{o}}\ell_j}[n_i^j]$).

Remark 5.2.10. It is convenient to think of an element $\underline{p} = (n^1, \ldots, (n_i^k)) \in \mathbf{N}^{\hat{\mathrm{o}}k}[t]$ as describing a family of planar rooted trees (see, e.g. [22]) with t leaves and k levels. More precisely, the numbers n_i^j describe the valence (number of input edges) to the i-th node at the j-th level, and a tree in this family is determined by a family of morphisms $\varphi_j \colon \mathbf{n^j} \to \mathbf{n^{j-1}}$ for $1 \leq j < k$ such that $|\varphi_j^{-1}(i)| = n_i^j$ for all i, j.

Let $\underline{Q}$ be so that $\underline{p} \circ \underline{Q}$ is defined. From this perspective, a tree in the family corresponding to $\underline{p} \circ \underline{Q}$ is build by "blowing up" each node n_i^j from $\underline{p}$ by a tree from the family corresponding to the profile $\underline{q}_i^j$ from $\underline{Q}$.

Definition 5.2.11. We define the *tensor* $\hat{\otimes}$ of reduced $\mathbf{N}_{\mathsf{lev}}$-objects $\mathcal{Q}^1, \cdots, \mathcal{Q}^k$ to be the left Kan extension of the following

$$
\coprod_{k \geq 0} \left(\mathbf{N}^{\hat{\mathrm{o}}k} \ltimes \left(\coprod_{\ell_1, \cdots, \ell_k \geq 0} \mathbf{N}^{\hat{\mathrm{o}}\ell_1} \times \cdots \times \mathbf{N}^{\hat{\mathrm{o}}\ell_k} \right) \right) \xrightarrow{\;\mathcal{Q}^1 \hat{\mathrm{o}} \cdots \hat{\mathrm{o}} \mathcal{Q}^k\;} \mathsf{C} \tag{5.2.12}
$$

$$
\downarrow {\scriptstyle (\underline{p}, \underline{Q}) \mapsto \underline{p} \circ \underline{Q}}
$$

$$
\coprod_{\ell \geq 0} \mathbf{N}^{\hat{\mathrm{o}}\ell} \xrightarrow[\text{left Kan ext.}]{\mathcal{Q}^1 \hat{\otimes} \cdots \hat{\otimes} \mathcal{Q}^k} \mathsf{C}
$$

such that if $\underline{p} \circ \underline{Q} \in \mathbf{N}^{\hat{\mathrm{o}}\ell_1 + \cdots + \ell_k}[t]$, then

$$
(\mathcal{Q}^1 \hat{\mathrm{o}} \cdots \hat{\mathrm{o}} \mathcal{Q}^k)(\underline{p} \circ \underline{Q}; t) := \mathcal{Q}_{\ell_1}^1(\underline{q}^1; n^1) \otimes \bigotimes_{i \in \mathbf{n^1}} \mathcal{Q}_{\ell_2}^2(\underline{q}_i^2; n_i^2) \cdots \otimes \bigotimes_{i \in \mathbf{n^{k-1}}} \mathcal{Q}_{\ell_k}^k(\underline{q}_i^k; n_i^k).
$$

Note then that $(\mathcal{Q}^{\hat{\otimes}k})_\ell \cong \coprod_{\ell_1 + \cdots + \ell_k = \ell} \mathcal{Q}_{\ell_1} \hat{\mathrm{o}} \cdots \hat{\mathrm{o}} \mathcal{Q}_{\ell_k}$, more specifically:

$$
(\mathcal{Q}^{\hat{\otimes}k})_\ell(\underline{p}; t) \cong \coprod_{\ell_1 + \cdots + \ell_k = \ell} \coprod_{\underline{p} = \underline{p}' \circ \underline{Q}} (\mathcal{Q} \hat{\mathrm{o}} \cdots \hat{\mathrm{o}} \mathcal{Q})(\underline{p} \circ \underline{Q}; t) \tag{5.2.13}
$$

where we note that the summands ℓ_j are ordered.

Definition 5.2.14. Let $\mathcal{P}$ and $\mathcal{Q}$ be reduced $\mathbf{N}_{\mathsf{lev}}$-objects in C. Their *nonsymmetric composition product* $\odot$ is defined as the coend $\mathcal{P}_- \otimes_{\mathsf{N}} \mathcal{Q}^{\hat{\otimes}-}$ where N denotes the category of finite sets $\mathbf{n}$ for $n \geq 0$ with only identity morphisms. That is,

$$(\mathcal{P} \odot \mathcal{Q})_\ell \cong \coprod_{k \geq 0} \mathcal{P}_k \dot{\otimes} (\mathcal{Q}^{\hat{\otimes} k})_\ell.$$

We use the notation $\dot{\otimes}$ to designate the product $\mathcal{P}_k \dot{\otimes}(\mathcal{Q}_{\ell_1} \hat{\circ} \cdots \hat{\circ} \mathcal{Q}_{\ell_k})$ is evaluated at a profile $(\underline{p}; t)$ as follows

$$(\mathcal{P}_k \dot{\otimes}(\mathcal{Q}_{\ell_1} \hat{\circ} \cdots \hat{\circ} \mathcal{Q}_{\ell_k}))(\underline{p}; t) \cong \coprod_{\underline{p} = \underline{p}' \circ \underline{Q}} \mathcal{P}_k(\underline{p}'; s') \otimes (\mathcal{Q}^1 \hat{\circ} \cdots \hat{\circ} \mathcal{Q}^k)(\underline{p} \circ \underline{Q}; t)$$

where $\underline{p}' \in \mathbf{N}^{\hat{\circ} k}[s']$ and $\underline{Q}$ is a family $(\underline{q}_i^j)$ as in (5.2.9) with $\underline{q}_i^j \in \mathbf{N}^{\hat{\circ} \ell_j}[s_i^j]$.

We necessarily then have

$$\underline{p}' = \left(s^1, (s_1^2, \ldots, s_{s^1}^2), \ldots, (s_1^k, \ldots, s_{s^{k-1}}^k)\right)$$

and can further describe $\mathcal{P} \odot \mathcal{Q}$ as

$$(\mathcal{P} \odot \mathcal{Q})_\ell(\underline{p}; t) \cong \coprod_{\ell_1 + \cdots + \ell_k = \ell} \coprod_{\underline{p} = \underline{p}' \circ \underline{Q}} \mathcal{P}_k(\underline{p}'; s') \otimes \bigotimes_{j=1}^{k} \left(\bigotimes_{i \in \mathbf{n}^j} \mathcal{Q}_{\ell_j}(\underline{q}_i^j; n_i^j) \right) \tag{5.2.15}$$

Example 5.2.16. We will evaluate $(\mathcal{P} \odot \mathcal{Q})_3$ at

$$\underline{p} = (n, (k_i)_{i \in \mathbf{n}}, (t_j)_{j \in \mathbf{k}}) \in \mathbf{N}^{\hat{\circ} 3}[t]$$

for $\mathcal{P}, \mathcal{Q}$ reduced $\mathbf{N}_{\mathsf{lev}}$-objects. Set $k := k_1 + \cdots + k_n$, we observe

$$\left(\mathcal{P}_2 \dot{\otimes}(\mathcal{Q}_1 \hat{\circ} \mathcal{Q}_2)\right)(\underline{p}; t) = \coprod_{\underline{p}=(n,(\underline{q}_1 \amalg \cdots \amalg \underline{q}_n))} \mathcal{P}_2(n,(s_1,\ldots,s_n);t) \otimes \left(\mathcal{Q}_1(n;n) \otimes \bigotimes_{i \in \mathbf{n}} \mathcal{Q}_2(\underline{q}_i; s_i) \right)$$

where $\underline{q}_i \in \mathbf{N}^{\hat{\circ}2}[s_i]$.

Using the language of Remark 5.2.10, we think of the above as partitioning the set of nodes $(t_j)_{j \in \mathbf{k}}$ from $\underline{p}$ into n sets of size $k_1, \ldots, k_n$, e.g., by defining a map $\varphi \colon \mathbf{k} \to \mathbf{n}$ such that $|\varphi^{-1}(i)| = k_i$ for $i = 1, \ldots, n$. Such a partition determines n profiles $\underline{q}_i = (k_i, (t_j)_{j \in \varphi^{-1}(i)}) \in \mathbf{N}^{\hat{\circ}2}[s_i]$ for $i = 1, \ldots, n$ where necessarily s_i is the sum $\sum_{j \in \varphi^{-1}(i)} t_j$. This precisely determines all possible ways of expressing the family of trees associated to $\underline{p}$ by a "vertex blowup" of the form $\underline{p} = \underline{p}' \circ \mathcal{Q}$, where $\underline{p}' \in \mathbf{N}^{\hat{\circ}2}[t]$, $\underline{q}^1 \in \mathbf{N}^{\hat{\circ}1}[n]$ and each $\underline{q}_i^2 \in \mathbf{N}^{\hat{\circ}2}$. The term $\left(\mathcal{P}_2 \dot{\otimes}(\mathcal{Q}_1 \hat{\circ} \mathcal{Q}_2)\right)(\underline{p}; t)$ is then obtained by using $\mathcal{P}$ to evaluate $\underline{p}'$ and $\mathcal{Q}$ to evaluate the profiles from $\underline{Q}$.

Similarly,

$$\left(\mathcal{P}_2 \dot{\otimes}(\mathcal{Q}_2 \hat{\circ} \mathcal{Q}_1)\right)(\underline{p}; t) = \mathcal{P}_2\left(k, (t_j)_{j \in \mathbf{k}}; t\right) \otimes \left(\mathcal{Q}_2(n, (k_i)_{i \in \mathbf{n}}; k) \otimes \bigotimes_{j \in \mathbf{k}} \mathcal{Q}_1(t_j; t_j) \right),$$

$$\left(\mathcal{P}_1 \dot{\otimes} \mathcal{Q}_3\right)(\underline{p}; t) = \mathcal{P}_1(t; t) \otimes \mathcal{Q}_3(\underline{p}; t),$$

$$\left(\mathcal{P}_3 \dot{\otimes}(\mathcal{Q}_1 \hat{\circ} \mathcal{Q}_1 \hat{\circ} \mathcal{Q}_1)\right)(\underline{p}; t) = \mathcal{P}_3(\underline{p}; t) \otimes \left(\mathcal{Q}_1(n;n) \otimes \bigotimes_{i \in \mathbf{n}} \mathcal{Q}_1(k_i; k_i) \otimes \bigotimes_{j \in \mathbf{k}} \mathcal{Q}_1(t_j; t_j) \right).$$

Proposition 5.2.17. *The category of $\mathbf{N}_{\mathsf{lev}}$-objects equipped with the composition product $\odot$ is monoidal.*

Proof. It is straightforward to verify that $\odot$ has a two-sided unit, $\mathcal{I}$, given by $\mathcal{I}_1(n;n) = \mathbf{1}$ and $\mathcal{I} = \phi$ otherwise. For $\mathbf{N}_{\mathsf{lev}}$-objects $\mathcal{P}, \mathcal{Q}, \mathcal{R}$, there is a natural isomorphism $(\mathcal{P} \odot \mathcal{Q}) \odot \mathcal{R} \cong$

$\mathcal{P}\odot(\mathcal{Q}\odot\mathcal{R})$ induced by the natural isomorphisms

$$\left(\mathcal{P}_n\dot\otimes\left(\mathcal{Q}_{k_1}\hat\circ\cdots\hat\circ\mathcal{Q}_{k_n}\right)\right)\dot\otimes\left(\mathcal{R}_{\ell_{1,1}}\hat\circ\cdots\hat\circ\mathcal{R}_{\ell_{n,k_n}}\right) \tag{5.2.18}$$

$$\cong \mathcal{P}_n\dot\otimes\left(\left(\mathcal{Q}_{k_1}\dot\otimes(\mathcal{R}_{\ell_{1,1}}\hat\circ\cdots\hat\circ\mathcal{R}_{\ell_{1,k_1}})\right)\hat\circ\cdots\hat\circ\left(\mathcal{Q}_{k_n}\dot\otimes(\mathcal{R}_{\ell_{n,1}}\hat\circ\cdots\hat\circ\mathcal{R}_{\ell_{n,k_n}})\right)\right)$$

obtained by a tedious but ultimately straightforward calculation. The remainder of the monoidal category axioms follow from similar observations. $\square$

Definition 5.2.19. A *nonsymmetric* $\mathbf{N}_{\mathsf{lev}}$-*operad* is a reduced $\mathbf{N}_{\mathsf{lev}}$-object $\mathcal{P}$ which is a monoid with respect to $\odot$. That is, there are unital and associative maps of $\mathbf{N}_{\mathsf{lev}}$-objects $\xi\colon \mathcal{P}\odot\mathcal{P}\to\mathcal{P}$ and $\varepsilon\colon\mathcal{I}\to\mathcal{P}$, i.e., such that the following diagrams commute

$$
\begin{array}{ccc}
\mathcal{P}\odot\mathcal{P}\odot\mathcal{P} & \xrightarrow{\ \xi\odot\mathrm{id}\ } & \mathcal{P}\odot\mathcal{P} \\
{\scriptstyle \mathrm{id}\odot\xi}\downarrow & & \downarrow{\scriptstyle \xi} \\
\mathcal{P}\odot\mathcal{P} & \xrightarrow{\ \xi\ } & \mathcal{P}
\end{array}
\qquad
\begin{array}{ccc}
\mathcal{P}\odot\mathcal{I} & \xrightarrow{\ \mathrm{id}\odot\varepsilon\ } & \mathcal{P}\odot\mathcal{P} & \xleftarrow{\ \varepsilon\odot\mathrm{id}\ } & \mathcal{I}\odot\mathcal{P} \\
& {\scriptstyle \cong}\searrow & \downarrow{\scriptstyle \xi} & \swarrow{\scriptstyle \cong} & \\
& & \mathcal{P} & &
\end{array}
$$

5.2.20   Algebras over a nonsymmetric $\mathbf{N}_{\mathsf{lev}}$-operad

Let $(\widehat{-})$ denote the inclusion of $\mathbf{N}$-colored objects to $\mathbf{N}_{\mathsf{lev}}$-objects given by

$$\widehat{X}_0(\emptyset;n) = X[n] \text{ and } \widehat{X}_k = \phi \text{ for } k\geq 1.$$

Note that $\widehat{X}$ is not reduced, but a straightforward modification of Definition 5.2.11 provides that $\left(\widehat{X}^{\hat\otimes n}\right)_0\cong X^{\hat\otimes n}$ and $\left(\widehat{X}^{\hat\otimes n}\right)_k\cong\phi$ for $k\geq 1$. Similarly, $(\widehat{-})$ is left adjoint to Ev_0 which takes values in nonsymmetric sequences and is defined at an $\mathbf{N}_{\mathsf{lev}}$-object $\mathcal{P}$ as

$$(\mathrm{Ev}_0\mathcal{P})[n] := \mathcal{P}_0(\emptyset;n).$$

If $\mathcal{P}$ is a nonsymmetric $\mathbf{N}_{\mathsf{lev}}$-operad then $\mathcal{P}\odot\widehat{X}$ remains concentrated at level 0 and hence defines a monad on **N**-colored objects

$$\mathcal{P}\odot(-)\colon X \mapsto \mathrm{Ev}_0(\mathcal{P}\odot\widehat{X}).$$

Definition 5.2.21. We say that an **N**-colored object X is an *algebra* over an nonsymmetric $\mathbf{N}_{\mathsf{lev}}$-operad $\mathcal{P}$ if there is an action map

$$\mathcal{P}\odot(X) \xrightarrow{\mu} X$$

which is associative and unital in that the following diagrams commute.

$$
\begin{array}{ccc}
\mathcal{P}\odot\mathcal{P}\odot(X) \xrightarrow{\xi\odot\mathrm{id}} \mathcal{P}\odot(X) & \qquad & \mathcal{P}\odot(X) \xrightarrow{\mu} X \\
\;\;\downarrow{\scriptstyle\mathrm{id}\odot\mu} \qquad\qquad \downarrow{\scriptstyle\mu} & & \;\;\uparrow{\scriptstyle\varepsilon} \quad\nearrow{\scriptstyle\cong} \\
\mathcal{P}\odot(X) \xrightarrow{\;\;\mu\;\;} X & & \mathcal{I}\odot(X)
\end{array}
$$

We denote by $\mathsf{Alg}^{\omega}_{\mathsf{lev}}(\mathcal{P})$ the category of algebras over a nonsymmetric $\mathbf{N}_{\mathsf{lev}}$-operad $\mathcal{P}$ along with $\mathcal{P}$-action preserving maps. Note that an action map μ consists of pieces

$$\mu_k\colon \mathcal{P}_k\dot{\otimes}(X^{\hat{\odot} k}) \to X$$

for $k \geq 0$ and that $\mathsf{Alg}^{\omega}_{\mathsf{lev}}(\mathcal{P})$ is complete and cocomplete and moreover that limits are built in the underlying category of **N**-colored objects.

5.2.22   Change of $\mathbf{N}_{\mathsf{lev}}$-operads adjunction

Given a map of nonsymmetric $\mathbf{N}_{\mathsf{lev}}$-operads $\sigma\colon \mathcal{P} \to \mathcal{Q}$ and a $\mathcal{P}$-algebra X we define $\mathcal{Q}\odot_{\mathcal{P}}(X)$ by the reflexive coequalizer

$$\mathcal{Q}\odot_{\mathcal{P}}(X) := \mathrm{colim}\left(\mathcal{Q}\odot(X) \leftleftarrows \mathcal{Q}\odot\mathcal{P}\odot(X) \right).$$

The top map above is given by $\mathcal{P} \odot (X) \xrightarrow{\mu_{\mathcal{P}}} X$ and the bottom is induced by the composite

$$\mathcal{Q} \odot \mathcal{P} \xrightarrow{\mathrm{id} \odot \sigma} \mathcal{Q} \odot \mathcal{Q} \xrightarrow{\xi_{\mathcal{Q}}} \mathcal{Q}.$$

The resulting object $\mathcal{Q} \odot_{\mathcal{P}}(X)$ inherits a natural $\mathcal{Q}$ algebra structure and the construction fits into an adjunction as in the following proposition.

Proposition 5.2.23. *Given a map of nonsymmetric $\mathbf{N}_{\mathsf{lev}}$-operads $\mathcal{P} \xrightarrow{\sigma} \mathcal{Q}$ there is a change of nonsymmetric $\mathbf{N}_{\mathsf{lev}}$-operads adjunction*

$$\mathsf{Alg}^{\omega}_{\mathsf{lev}}(\mathcal{P}) \underset{\sigma^*}{\overset{\mathcal{Q} \odot_{\mathcal{P}}(-)}{\rightleftarrows}} \mathsf{Alg}^{\omega}_{\mathsf{lev}}(\mathcal{Q})$$

with right adjoint σ^* given by restriction along σ.

5.2.24   A forgetful functor to N-colored operads

We describe forgetful functor $\mathbb{U}$ from $\mathbf{N}_{\mathsf{lev}}$-operads to $\mathbf{N}$-colored operads (specifically, nonsymmetric $\mathbf{N}$-colored operads). Given $\underline{p} = ((n^1, \cdots, (n_i^{\ell})_{i \in \mathbf{n}^{\ell-1}}) \in \mathbf{N}^{\delta \ell}$, we set $s(\underline{p})$ to be the unordered list of the elements of the levels of $\underline{p}$, i.e.,

$$s(\underline{p}) := \left\{ n_i^j : j \in \{1, \cdots, n\}, i \in \mathbf{n}^{\mathbf{j}} \right\}.$$

Given an $\mathbf{N}_{\mathsf{lev}}$-object $\mathcal{Q}$ we define $\mathbb{U}\mathcal{Q}$ by

$$(\mathbb{U}\mathcal{Q})(c_1, \ldots, c_k; t) := \coprod_{s(\underline{p})=(c_1,\ldots,c_k)} \mathcal{Q}_{\ell}(\underline{p}; t) \tag{5.2.25}$$

where the coproduct ranges over $\underline{p} \in \coprod_{\ell \geq 0} \mathbf{N}^{\delta \ell}$. We leave the proof of the following proposition to the reader.

Proposition 5.2.26. *If $\mathcal{P}$ is an $\mathbf{N}_{\mathsf{lev}}$-operad then $\mathbb{U}\mathcal{P}$ is a (nonsymmetric) $\mathbf{N}$-colored operad.*

Furthermore, the categories $\mathsf{Alg}^{\omega}_{\mathsf{lev}}(\mathcal{P})$ and $\mathsf{Alg}_{\mathrm{U}\mathcal{P}}$ are equivalent.

5.3 Symmetric $\mathbf{N}_{\mathsf{lev}}$-objects

We now impart symmetric group actions on our $\mathbf{N}_{\mathsf{lev}}$-objects in a way that captures operadic composition. Denote by $\mathcal{I}^{\Sigma}$ the $\mathbf{N}_{\mathsf{lev}}$-object in $(\mathsf{C}, \otimes, \mathbf{1})$ with

$$
\mathcal{I}^{\Sigma}_{\ell}(\underline{p}; t) = \begin{cases} \boldsymbol{\Sigma}[n] & \ell = 1, \underline{p} = n = t \\ \phi & \text{otherwise} \end{cases}
$$

Recall here that $\boldsymbol{\Sigma}[n] = \coprod_{\sigma \in \Sigma_n} \mathbf{1}$. Note that $\mathcal{I}^{\Sigma}$ is a nonsymmetric $\mathbf{N}_{\mathsf{lev}}$-operad whose composition maps are induced by the block matrix inclusions

$$
\Sigma_n \times (\Sigma_{k_1} \times \cdots \times \Sigma_{k_n}) \to \Sigma_{k_1 + \cdots + k_n}.
$$

Moreover the data of an algebra over $\mathcal{I}^{\Sigma}$ is precisely that of a symmetric sequence; i.e., $\mathsf{Alg}^{\omega}_{\mathcal{I}^{\Sigma}} \cong \mathsf{SymSeq}$.

Definition 5.3.1. An $\mathbf{N}_{\mathsf{lev}}$-object $\mathcal{P}$ *symmetric* if $\mathcal{P}$ has compatible right and left actions of $\mathcal{I}^{\Sigma}$ in that the following diagram must commute

$$
\begin{array}{ccc}
\mathcal{I}^{\Sigma} \odot \mathcal{P} \odot \mathcal{I}^{\Sigma} & \xrightarrow{\mu_{\ell} \odot \mathrm{id}} & \mathcal{P} \odot \mathcal{I}^{\Sigma} \\
{\scriptstyle \mathrm{id} \odot \mu_r} \downarrow & & \downarrow {\scriptstyle \mu_r} \\
\mathcal{I}^{\Sigma} \odot \mathcal{P} & \xrightarrow{\mu_{\ell}} & \mathcal{P}
\end{array}
$$

where μ_{ℓ} (resp. μ_r) denotes the left (resp. right) action map of $\mathcal{I}^{\Sigma}$ on $\mathcal{P}$.

In other words, a symmetric $\mathbf{N}_{\mathsf{lev}}$-object is an $(\mathcal{I}^{\Sigma}, \mathcal{I}^{\Sigma})$-bimodule. Note that $\mathcal{I}^{\Sigma} \odot (X) \cong \Sigma \cdot X$ is the free symmetric sequence on X (see also Remark 5.4.14).

5.3.2 Symmetric $\mathbf{N}_{\mathsf{lev}}$-operads

Definition 5.3.3. Let $\mathcal{P}, \mathcal{Q}$ be $(\mathcal{I}^{\Sigma}, \mathcal{I}^{\Sigma})$-bimodules. We define their *symmetric composition product*, denoted $\mathcal{P} \odot_{\Sigma} \mathcal{Q}$, as the (reflexive) coequalizer (calculated in symmetric $\mathbf{N}_{\mathsf{lev}}$-objects)

$$\mathcal{P} \odot_{\Sigma} \mathcal{Q} := \mathcal{P} \odot_{\mathcal{I}^{\Sigma}} \mathcal{Q} \cong \operatorname{colim} \left(\mathcal{P} \odot \mathcal{Q} \rightleftarrows \mathcal{P} \odot \mathcal{I}^{\Sigma} \odot \mathcal{Q} \right)$$

where the two maps are induced by the left and right actions actions of $\mathcal{I}^{\Sigma}$ on $\mathcal{Q}$ and $\mathcal{P}$.

Note that $\mathcal{P} \odot_{\Sigma} \mathcal{Q}$ inherits left and right $\mathcal{I}^{\Sigma}$ actions by those on $\mathcal{P}$ and $\mathcal{Q}$ respectively, and so remains an $(\mathcal{I}^{\Sigma}, \mathcal{I}^{\Sigma})$-bimodule. Moreover, $\mathcal{I}^{\Sigma}$ is a two-sided unit for $\odot_{\Sigma}$ and symmetric $\mathbf{N}_{\mathsf{lev}}$-objects equipped with the product $(\odot_{\Sigma}, \mathcal{I}^{\Sigma})$ is a monoidal category.

Remark 5.3.4. Since $\mathcal{I}^{\Sigma}$ is concentrated at level 1, is it possible to further describe the object $\mathcal{P} \odot_{\Sigma} \mathcal{Q}$ in terms of its constituent parts. In particular,

$$(\mathcal{P} \odot_{\Sigma} \mathcal{Q})_{\ell} \cong \coprod_{k \geq 0} \coprod_{\ell_1 + \cdots + \ell_k = \ell} \mathcal{P}_k \dot{\otimes}_{\Sigma} (\mathcal{Q}_{\ell_1} \hat{\mathrm{o}} \cdots \hat{\mathrm{o}} \mathcal{Q}_{\ell_k})$$

where $\mathcal{P}_k \dot{\otimes}_{\Sigma} (\mathcal{Q}_{\ell_1} \hat{\mathrm{o}} \cdots \hat{\mathrm{o}} \mathcal{Q}_{\ell_k})$ is obtained as the coequalizer

$$\operatorname{colim} \left(\mathcal{P}_k \dot{\otimes} (\mathcal{Q}_{\ell_1} \hat{\mathrm{o}} \cdots \hat{\mathrm{o}} \mathcal{Q}_{\ell_k}) \rightleftarrows \left(\mathcal{P}_k \dot{\otimes} (\underbrace{\mathcal{I}_1^{\Sigma} \hat{\mathrm{o}} \cdots \hat{\mathrm{o}} \mathcal{I}_1^{\Sigma}}_{k}) \right) \dot{\otimes} (\mathcal{Q}_{\ell_1} \hat{\mathrm{o}} \cdots \hat{\mathrm{o}} \mathcal{Q}_{\ell_k}) \right)$$

such that the top is induced by the right action of $\mathcal{I}^{\Sigma}$ on $\mathcal{P}$ and the bottom map is induced by the isomorphism (5.2.18) and the left action of $\mathcal{I}^{\Sigma}$ on $\mathcal{Q}$.

Definition 5.3.5. A *symmetric $\mathbf{N}_{\mathsf{lev}}$-operad* is a reduced symmetric $\mathbf{N}_{\mathsf{lev}}$-object $\mathcal{P}$, which is a monoid with respect to $\odot_{\Sigma}$. That is, there is a multiplication map $\xi \colon \mathcal{P} \odot_{\Sigma} \mathcal{P} \to \mathcal{P}$ and unit map $\varepsilon \colon \mathcal{I}^{\Sigma} \to \mathcal{P}$ that satisfy the usual associativity and unitality conditions.

5.3.6 Algebras over symmetric $\mathbf{N}_{\mathsf{lev}}$-operads

We now define an algebra over a symmetric $\mathbf{N}_{\mathsf{lev}}$-operad $\mathcal{P}$. Note than algebra over a symmetric $\mathbf{N}_{\mathsf{lev}}$-operad is a symmetric $\mathbf{N}_{\mathsf{lev}}$-object concentrated at level 0, that is, an $\mathcal{I}^{\Sigma}$-algebra or symmetric sequence. As before, given a symmetric $\mathbf{N}_{\mathsf{lev}}$-operad $\mathcal{P}$, let

$$\mathcal{P}\odot_{\Sigma}(-)\colon X \mapsto \mathrm{Ev}_0(\mathcal{P}\odot_{\Sigma}\widehat{X})$$

be the associated monad on SymSeq.

Definition 5.3.7. A symmetric sequence X is an *algebra* over a symmetric $\mathbf{N}_{\mathsf{lev}}$-operad $\mathcal{P}$ if there is an action map $\mu\colon \mathcal{P}\odot_{\Sigma}(X) \to X$ which is associative and unital (as in Definition 5.2.21 with $\odot$ replaced by $\odot_{\Sigma}$ q.v.)

We denote by $\mathsf{Alg}^{\Sigma}_{\mathsf{lev}}(\mathcal{P})$ the category of symmetric $\mathcal{P}$-algebras with $\mathcal{P}$-algebra preserving maps; for simplicity we will frequently use $\mathsf{Alg}_{\mathcal{P}}$ instead when there is no room for confusion. We note that μ consists of maps

$$\mu_k\colon \mathcal{P}_k\dot{\otimes}_{\Sigma}(X^{\hat{\delta}k}) \to X$$

where the action of $\mathcal{I}^{\Sigma}$ on $X^{\hat{\delta}k}$ agrees with that for symmetric sequences discussed in Section 2.1. Furthermore,

$$\mu_0\colon I \cong \mathcal{P}_0\dot{\otimes}_{\Sigma}(X^{\hat{\delta}0}) \to X$$

gives a unit map for $X \in \mathsf{Alg}_{\mathcal{P}}$ and we note that an algebra X over $\mathcal{P}$ will always be reduced, i.e., $X[0] = \phi$.

Example 5.3.8 (Free symmetric $\mathcal{P}$-algebra on a symmetric sequence). Given a symmetric sequence X, the object $\mathcal{P}\odot_{\Sigma}(X)$ is the free $\mathcal{P}$-algebra on X and fits into an adjunction

$$\mathsf{SymSeq}_{\mathsf{C}} \underset{\mathcal{U}}{\overset{\mathcal{P}\odot_{\Sigma}(-)}{\rightleftarrows}} \mathsf{Alg}_{\mathcal{P}}$$

where $\mathcal{U}$ is the forgetful functor. In particular, $\mathsf{Oper}\odot_\Sigma(X)$ (see Definition 5.4.2) is the *free operad* on X (see, e.g., [2, 9.4]).

We leave the proof of the following to the reader as it follows from standard arguments as in [44, 3.29] or [18, 4.3].

Proposition 5.3.9. *If $(\mathsf{C}, \otimes, \mathbf{1})$ is closed symmetric monoidal which contains all small limits and colimits, then all small limits and colimits exist in $\mathsf{Alg}_\mathcal{P}$. Limits and filtered colimits are built in the underlying category of symmetric sequences and are further reflected by the forgetful functor $\mathcal{U}$.*

General colimits shaped on a small diagram $\mathcal{D}$ are constructed by the following (reflexive) coequalizer (whose colimits are constructed in SymSeq):

$$\operatorname{colim}_{d \in \mathcal{D}} X_d \cong \operatorname{colim} \left(\mathcal{P} \odot_\Sigma (\operatorname{colim}_{d \in \mathcal{D}} X_d) \Longleftarrow \mathcal{P} \odot_\Sigma (\operatorname{colim}_{d \in \mathcal{D}} \mathcal{P} \odot_\Sigma (X_d)) \right).$$

5.3.10 Modules over $\mathcal{P}$-algebras

Definition 5.3.11. Let $\mathcal{P}$ be a symmetric $\mathbf{N}_{\mathsf{lev}}$-operad and $\mathcal{W}$ be a $\mathcal{P}$-algebra. Let M be a symmetric sequence. We say that M is an $\mathcal{W}$-*module* if there are maps of the form

$$\eta_\ell \colon \mathcal{P}_\ell \dot{\otimes}_\Sigma \left(\mathcal{W}^{\hat{\circ}(\ell-1)} \hat{\circ} M \right) \to M$$

for $\ell \geq 1$ that satisfy associativity (5.3.12) and unitality (5.3.13). If M is concentrated at level 0 we say that the object $M[0]$ is a $\mathcal{W}$-*algebra*.

Set $\xi \colon \mathcal{P} \odot_\Sigma \mathcal{P} \to \mathcal{P}$ to be the multiplication on $\mathcal{P}$ and $\mu \colon \mathcal{P} \odot_\Sigma (\mathcal{W}) \to \mathcal{W}$ the action map on $\mathcal{W}$. Let $\ell := \ell_1 + \cdots + \ell_k$. Associativity and unitality amounts to the commutitivity of the following diagrams

$$
\begin{array}{ccc}
\left(\mathcal{P}_k\dot{\otimes}_\Sigma(\mathcal{P}_{\ell_1}\hat{\circ}\cdots\hat{\circ}\mathcal{P}_{\ell_k})\right)\dot{\otimes}_\Sigma\left(\mathcal{W}^{\hat{\circ}(\ell-1)}\hat{\circ}M\right) & \xrightarrow{\;\xi_\ell\dot{\otimes}_\Sigma\mathrm{id}\;} & \mathcal{P}_\ell\dot{\otimes}_\Sigma\left(\mathcal{W}^{\hat{\circ}(\ell-1)}\hat{\circ}M\right) \\
\downarrow{\scriptstyle\cong} & & \\
\mathcal{P}_k\dot{\otimes}_\Sigma\left((\mathcal{P}_{\ell_1}\dot{\otimes}_\Sigma\mathcal{W}^{\hat{\circ}\ell_1})\hat{\circ}\cdots\hat{\circ}(\mathcal{P}_{\ell_k}\dot{\otimes}_\Sigma(\mathcal{W}^{\hat{\circ}\ell_k-1}\hat{\circ}M)))\right) & & \downarrow{\scriptstyle\eta_\ell} \\
\downarrow{\scriptstyle\mathrm{id}\dot{\otimes}_\Sigma(\mu_{\ell_1}\hat{\circ}\cdots\hat{\circ}\mu_{\ell_{k-1}}\hat{\circ}\eta_{\ell_k})} & & \\
\mathcal{P}_k\dot{\otimes}_\Sigma\left(\underbrace{\mathcal{W}\hat{\circ}\cdots\hat{\circ}\mathcal{W}}_{k-1}\hat{\circ}M\right) & \xrightarrow{\;\eta_k\;} & M,
\end{array}
\tag{5.3.12}
$$

and

$$
\begin{array}{ccc}
\mathcal{P}_2\dot{\otimes}_\Sigma\left((\mathcal{P}_0\dot{\otimes}_\Sigma\mathcal{W}^{\hat{\circ}0})\hat{\circ}(\mathcal{P}_1\dot{\otimes}_\Sigma M)\right) & \xrightarrow{\;\mathrm{id}\dot{\otimes}_\Sigma(\mu_0\hat{\circ}\eta_1)\;} & \mathcal{P}_2\dot{\otimes}_\Sigma(\mathcal{W}\hat{\circ}M) \\
\downarrow{\scriptstyle\cong} & & \\
\left(\mathcal{P}_2\dot{\otimes}_\Sigma(\mathcal{P}_0\hat{\circ}\mathcal{P}_1)\right)\dot{\otimes}_\Sigma M & & \downarrow{\scriptstyle\eta_2} \\
\downarrow{\scriptstyle\xi_1\dot{\otimes}_\Sigma\mathrm{id}} & & \\
\mathcal{P}_1\dot{\otimes}_\Sigma M & \xrightarrow{\;\eta_1\;} & M.
\end{array}
\tag{5.3.13}
$$

Recall that $\mu_0\colon I\cong\mathcal{P}_0\dot{\otimes}_\Sigma\mathcal{W}^{\hat{\circ}0}\to\mathcal{W}$ is the unit map for $\mathcal{W}$.

Remark 5.3.14. We encourage the reader to compare the above definition with that of modules over algebras over an operad, e.g., as in May [61, Definition 3]. In [13, 1.5.1] an example of a 2-colored operad whose algebras are pairs (A, M) of an $\mathcal{O}$-algebra A along with an A-module M is provided. The pair $(\mathcal{W}, M)$ can be described analogously as an algebra over an $\mathbf{N}_+ := \{*, 0, 1, 2, \dots\}$-colored operad with levels, though we will not require such description.

Definition 5.3.15. We say a map $\mathcal{P}\to\mathcal{Q}$ of (symmetric) $\mathbf{N}_{\mathsf{lev}}$-operads in some symmetric monoidal model category C is an *equivalence* if for any $\underline{p}\in\mathbf{N}^{\hat{\circ}k}[t]$ the induced map $\mathcal{P}_k(\underline{p};t)\to\mathcal{Q}_k(\underline{p};t)$ is a weak equivalence in C. We write $\mathcal{P}\simeq\mathcal{Q}$ if there is a zig-zag of equivalences of (symmetric) $\mathbf{N}_{\mathsf{lev}}$-operads connecting $\mathcal{P}$ and $\mathcal{Q}$.

In the special case that $\mathcal{P} \simeq \mathsf{Oper}$ then we say that a $\mathcal{P}$-algebra $\mathcal{W}$ is an A_∞-*operad* and that modules over $\mathcal{W}$ are A_∞-*algebras*.

5.4 Examples of symmetric $\mathbf{N}_{\mathsf{lev}}$-operads

In this section we describe some examples of symmetric $\mathbf{N}_{\mathsf{lev}}$-operads of interest, specifically the coendomorphism $\mathbf{N}_{\mathsf{lev}}$-operads on a given cosimplicial symmetric sequence. We begin by describing Oper – the symmetric $\mathbf{N}_{\mathsf{lev}}$-operad whose algebras are (one-color) operads as some of its properties will be essential in what is to come. Our eventual goal is to prove that the coendomorphism $\mathbf{N}_{\mathsf{lev}}$-operad on a Σ-copowered symmetric sequence $\mathcal{X}$ (see Remark 5.4.14) is indeed a symmetric $\mathbf{N}_{\mathsf{lev}}$-operad; with the particular example of $\mathcal{A} = \mathsf{coEnd}(\Sigma \cdot \Delta_+^\bullet)$ in mind (see Section 5.4.25).

Though we write most of this section for a general closed cocomplete symmetric monoidal category C, we invite the reader to think particularly of the cases when $\mathsf{C} = \mathsf{Spt}$ or Top_*.

5.4.1 The symmetric $\mathbf{N}_{\mathsf{lev}}$-operad Oper

We begin by describing Oper for the category Set of sets.

Definition 5.4.2. Let Σ denote the symmetric sequence in $(\mathsf{Set}, \times, *)$ with $\Sigma[n] = \Sigma_n$ and define a reduced $\mathbf{N}_{\mathsf{lev}}$-object as follows. For $\underline{p} \in \mathbf{N}^{\hat{o}k}[t]$ we set

$$\mathsf{Oper}_\ell(\underline{p}; t) := \hom\left(\Sigma[t], \Sigma^{\square \ell}[\underline{p}]\right)^{\Sigma_t}$$

Remark 5.4.3. Note there are isomorphisms

$$\mathsf{Oper}_\ell(\underline{p}; t) = \hom\left(\Sigma[t], \Sigma^{\circ \ell}[\underline{p}]\right)^{\Sigma_t} \cong \hom\left(*, \Sigma^{\circ \ell}[\underline{p}]\right) \cong \Sigma^{\circ \ell}[\underline{p}]. \tag{5.4.4}$$

Computing some small examples of Oper, we note that

$$\mathsf{Oper}_0(\emptyset; 1) \cong * \qquad\qquad \mathsf{Oper}_1(\emptyset; n) \cong \emptyset \quad (n \neq 1)$$

$$\mathsf{Oper}_1(n; n) \cong \Sigma_n \quad (n \geq 0) \qquad\qquad \mathsf{Oper}_1(n; m) \cong \emptyset \quad (n \neq m \geq 0)$$

$$\mathsf{Oper}_2(n, (k_1, \ldots, k_n); k) \cong \Sigma_n \times_{\Sigma_{p_1} \times \cdots \times \Sigma_{p_n}} \Sigma_k$$

where $p_1, \ldots, p_m$ denotes the multiplicities of distinct integers among $k_1, \ldots, k_n$, $k = \sum_{i=1}^n k_i$, and $\Sigma_{p_1} \times \cdots \times \Sigma_{p_m}$ acts on Σ_k, e.g., by permutation of block matrices

$$\Sigma_{k_1} \times \cdots \times \Sigma_{k_n} \leq \Sigma_k.$$

Similarly, let $q_1, \ldots, q_r$ denotes the multiplicities of the distinct integers among $t_1, \ldots, t_k$ and set

$$\underline{p} = (n, (k_1, \ldots, k_n), (t_1, \ldots, t_k)) \in \mathbf{N}^{\hat{o}3}[t].$$

Then

$$\mathsf{Oper}_3(\underline{p}; t) \cong \Sigma_n \times_{\Sigma_{p_1} \times \cdots \times \Sigma_{p_m}} \Sigma_k \times_{\Sigma_{q_1} \times \cdots \times \Sigma_{q_r}} \Sigma_t$$

Proposition 5.4.5. Oper *is a symmetric* $\mathbf{N}_{\mathsf{lev}}$*-operad.*

Proof. As we will see, Oper is particularly special as the structure maps

$$\xi_{k,(\ell_1,\ldots,\ell_k)} \colon \mathsf{Oper}_k \hat{\otimes}_\Sigma (\mathsf{Oper}_{\ell_1} \hat{o} \cdots \hat{o} \mathsf{Oper}_{\ell_k}) \to \mathsf{Oper}_\ell \tag{5.4.6}$$

which comprise $\xi \colon \mathsf{Oper} \odot_\Sigma \mathsf{Oper} \to \mathsf{Oper}$ consist of isomorphisms once evaluated at a profile $\underline{p} \in \mathbf{N}^{\hat{o}\ell}[t]$.

That Oper is symmetric follows from the first part of the proof of Proposition 5.4.15.

The unit map $\epsilon\colon \mathcal{I}^{\Sigma} \to \mathsf{Oper}$ is obtained via the identity morphisms

$$\mathcal{I}_1^{\Sigma}(n;n) \cong \Sigma_n \to \Sigma_n \cong \mathsf{Oper}_1(n;n)$$

and the initial morphism elsewhere. Let us now produce the desired map (5.4.6) at a profile $\underline{p} \in \mathbf{N}^{\hat{\circ}\ell}[t]$.

For the reader who finds the following constructions a bit opaque, we first provide the following intuition: for $\ell \geq 0$ let $\circ_\ell\colon \mathsf{SymSeq}^{\times \ell} \to \mathsf{SymSeq}$ be the functor $\circ_\ell(X_1, \ldots, X_\ell) = X_1 \circ \cdots \circ X_\ell$. Since $\circ$ is strictly monoidal, there are *isomorphisms*

$$\Xi_{k,(\ell_1,\cdots,\ell_k)}\colon \ \circ_k\left(\circ_{\ell_1}, \cdots, \circ_{\ell_k}\right) \xrightarrow{\cong} \circ_{\ell_1 + \cdots + \ell_k} \tag{5.4.7}$$

such that $\circ_\bullet$ is a *nonsymmetric functor-operad* (see, e.g., McClure-Smith [65, §4], omitting the requirement of symmetric group actions). Moreover, the composition maps $\xi_{k,(\ell_1,\ldots,\ell_k)}$ are precisely the morphisms which prescribe the equivariance of the isomorphism $\Xi_{k,(\ell_1,\ldots,\ell_k)}$ once evaluated at a particular string of inputs, given that evaluation at a profile in $\mathbf{N}^{\hat{\circ}\ell}$ is the same as evaluating a symmetric sequence *from the left*. For instance, $\xi_{3,(2,1,3)}$ provides the isomorphisms (natural in $X_1, \ldots, X_6$)

$$(X_1 \circ X_2) \circ X_3 \circ (X_4 \circ X_5 \circ X_6) \cong X_1 \circ \cdots \circ X_6$$

and moreover, given $\underline{p} \in \mathbf{N}^{\hat{\circ}\ell}[t]$, the desired map $\xi_{k,(\ell_1,\ldots,\ell_k)}[\underline{p}]$ may be thought of a precisely arising from the isomorphism

$$\left(\Sigma^{\circ \ell_1} \circ \cdots \circ \Sigma^{\circ \ell_k}\right)[\underline{p}] \xrightarrow{\cong} \Sigma^{\circ \ell}[\underline{p}].$$

We describe $\xi_{2,(1,2)}$ first and note the general case follows a similar argument. Let $\underline{p} \in$

$\mathbf{N}^{\hat{\circ}3}[t]$ and note that

$$\mathsf{Oper}_2 \dot{\otimes}_\Sigma (\mathsf{Oper}_1 \hat{\circ} \mathsf{Oper}_2))[\underline{p}] \cong \coprod_{\underline{p}=(n,(\underline{p}_1 \amalg \cdots \amalg \underline{p}_n))} \mathsf{Oper}_2 \dot{\otimes}_\Sigma (\mathsf{Oper}_1 \hat{\circ} \mathsf{Oper}_2))[(n, (\underline{p}_1, \ldots, \underline{p}_n))].$$

Fix $\underline{p}_i \in \mathbf{N}^{\hat{\circ}2}[s_i]$ for $i = 1, \ldots, n$ such that $\underline{p} = (n, (\underline{p_1}\amalg\cdots\amalg\underline{p_n}))$ and set $\underline{p}' = (n, (s_1, \ldots, s_n)) \in \mathbf{N}^{\hat{\circ}2}[t]$. We then observe

$$(\mathsf{Oper}_2 \dot{\otimes}_\Sigma (\mathsf{Oper}_1 \hat{\circ} \mathsf{Oper}_2))[(n, (\underline{p}_1, \ldots, \underline{p}_n))] \tag{5.4.8}$$
$$\cong \Sigma^{\circ 2}[\underline{p}'] \times_{S(\underline{p}')} \left(\Sigma[n] \times (\Sigma^{\circ 2}[\underline{p}_1] \times \cdots \times \Sigma^{\circ 2}[\underline{p}_n]) \right)$$
$$\cong \Sigma^{\circ 3}[(n, (\underline{p}_1, \cdots, \underline{p}_n))] \xrightarrow{\iota} \Sigma^{\circ 3}[\underline{p}] \cong \mathsf{Oper}_3(\underline{p}; t)$$

such that $S(\underline{p}') = \Sigma_n \times \prod_{i=1}^n \Sigma_{s_i}$ and ι is the natural inclusion obtained from the assumption $\underline{p} = (n, (\underline{p}_1 \amalg \cdots \amalg \underline{p}_n))$.

The desired map $\xi_{2,(1,2)}[\underline{p}]$ is induced by the coproduct of composites (5.4.8) for all $\underline{p} = (n, (\underline{p}_1, \ldots, \underline{p}_n))$. Note further that *as sets* there is an isomorphism

$$\coprod_{\underline{p}=(n,(\underline{p}_1 \amalg \cdots \amalg \underline{p}_n))} \Sigma^{\circ 3}[(n, (\underline{p}_1, \cdots, \underline{p}_n))] \cong \Sigma^{\circ 3}[\underline{p}]$$

since $\circ$ is strictly monoidal in the category of symmetric sequences of sets. Thus, $\xi_{2,(1,2)}[\underline{p}]$ is invertible and more generally $\xi_{k,(\ell_1,\ldots,\ell_k)}$ evaluated at any profile in $\mathbf{N}^{\hat{\circ}\ell}$ is also invertible.

Associativity of ξ then follows from the associativity of Ξ as in (5.4.7). That is, for a profile $(n, (k_1, \ldots, k_n)) \in \mathbf{N}^{\hat{\circ}2}[k]$ and for $i = 1, \ldots, n$, $\underline{q}_i = (k_i, (\ell_{i,1}, \cdots, \ell_{i,k_i})) \in \mathbf{N}^{\hat{\circ}2}[t_i]$ the associativity relation

$$\xi_{k,(\ell_{1,1}, \cdots, \ell_{n,k_n})} \left(\xi_{n,(k_1,\cdots,k_n)} \dot{\otimes}_\Sigma \mathrm{id} \right)$$
$$= \xi_{n,(t_1,\cdots,t_n)} \left(\mathrm{id} \dot{\otimes}_\Sigma (\xi_{k_1,(\ell_{1,1},\cdots,\ell_{1,k_1})} \hat{\circ} \cdots \hat{\circ} \xi_{k_n,(\ell_{n,1},\cdots,\ell_{n,k_n})}) \right)$$

evaluated at some $\underline{p} \in \mathbf{N}^{\hat{o}\ell}$ follows from the commutative square of isomorphisms

$$\left(\left(\Sigma^{\circ\ell_{1,1}} \circ \cdots \circ \Sigma^{\circ\ell_{1,k_1}}\right) \circ \cdots \circ \left(\Sigma^{\circ\ell_{n,1}} \circ \cdots \circ \Sigma^{\circ\ell_{n,k_n}}\right)\right)[\underline{p}] \longrightarrow \left(\Sigma^{\circ t_1} \circ \cdots \circ \Sigma^{\circ t_n}\right)[\underline{p}]$$

$$\downarrow \qquad\qquad\qquad\qquad\qquad\qquad\qquad\qquad\qquad\qquad \downarrow$$

$$\left(\Sigma^{\circ\ell_{1,1}} \circ \cdots \circ \Sigma^{\circ\ell_{n,k_n}}\right)[\underline{p}] \longrightarrow \Sigma^{\circ\ell}[\underline{p}]$$

Similarly, the unitality condition is satisfied by the more obvious isomorphisms

$$\left(\underbrace{(\Sigma \circ \cdots \circ \Sigma)}_{n}\right)[\underline{p}] \cong \Sigma^{\circ n}[\underline{p}] \cong \left(\underbrace{(\Sigma) \circ \cdots \circ (\Sigma)}_{n}\right)[\underline{p}]$$

for all $n \geq 0$ and $\underline{p} \in \mathbf{N}^{\hat{o}1}$ (i.e., $\underline{p} = p \geq 0$). $\qquad\qquad\qquad\qquad\square$

Remark 5.4.9. Let $(\mathsf{C}, \otimes, \mathbf{1})$ be a closed symmetric monoidal category with finite coproducts. We write $\mathsf{Oper}^{\mathsf{C}}$ for the image of Oper in C under $\Sigma_n \mapsto \Sigma[n] \cong \coprod_{\sigma \in \Sigma_n} \mathbf{1}$. That is, given a profile $\underline{p} \in \mathbf{N}^{\hat{o}k}[t]$ we set

$$\mathsf{Oper}^{\mathsf{C}}(\underline{p}; t) = \mathrm{Map}^{\mathsf{C}}\left(\Sigma[t], \Sigma^{\circ k}[\underline{p}]\right)^{\Sigma_t}.$$

Before showing that Oper encodes (one-color) operads as its algebras we first demonstrate another class of symmetric $\mathbf{N}_{\mathsf{lev}}$-operads.

5.4.10   Coendomorphism symmetric $\mathbf{N}_{\mathsf{lev}}$-operads

Recall as in Section 6 that $(\mathsf{C}, \otimes, \mathbf{1})$ denotes a closed cocomplete symmetric monoidal category and Σ is the symmetric sequence in C with $\Sigma[k] = \coprod_{\sigma \in \Sigma_k} \mathbf{1}$.

Definition 5.4.11. Let $\mathcal{X} \in \mathsf{SymSeq}_{\mathsf{C}}^{\Delta}$ and set $\mathrm{coEnd}(\mathcal{X})$ to be the reduced $\mathbf{N}_{\mathsf{lev}}$-object given at $(\underline{p}; t) \in \mathbf{N}^{\hat{o}\ell}[t]$ by

$$\mathsf{coEnd}(\mathcal{X})_\ell(\underline{p}; t) := \mathrm{Map}_{\mathbf{\Delta}^{\mathrm{res}}}\left(\mathcal{X}[t], \mathcal{X}^{\mathring{\Box}\ell}[\underline{p}]\right)^{\Sigma_t}.$$

Example 5.4.12. Unravelling the above definition, $\mathsf{coEnd}(\mathcal{X})_1(k; k)$ consists of all Σ_k-equivariant cosimplicial maps $\mathcal{X}[k] \to \mathcal{X}[k]$. Let $(\underline{q}; k) = (n, (k_1, \ldots, k_n); k) \in \mathbf{N}^{\hat{o}2}[k]$ and recall the description of $H(k_1, \ldots, k_n) \leq \Sigma_k$ from Definition 2.1.3. Then, $\mathsf{coEnd}(\mathcal{X})_2(\underline{q}; k)$ consists of all Σ_k-equivariant cosimplicial maps of the form

$$\mathcal{X}[k] \to (\mathcal{X}\mathring{\Box}\mathcal{X})[\underline{q}] \cong \Sigma[k] \otimes_{H(k_1, \ldots, k_n)} \mathcal{X}[n]\Box(\mathcal{X}[k_1] \otimes \cdots \otimes \mathcal{X}[k_n]).$$

Further, $\mathsf{coEnd}(\mathcal{X})$ is *quadratic* in that it is generated by its first two levels as follows: let $\underline{p} = (n, (k_i)_{i\in\mathbf{n}}, (t_j)_{j\in\mathbf{k}})$ and set $k := \sum_{i=1}^n k_i$ and $t := \sum_{j=1}^k t_j$. Then, $\mathsf{coEnd}(\mathcal{X})_3(\underline{p}; t)$ consists of cosimplicial maps ψ that fit into the following Σ_t-equivariant diagram

$$
\begin{array}{ccc}
\mathcal{X}[t] & \xrightarrow{\;\;\psi_1\;\;} & (\mathcal{X}\mathring{\Box}\mathcal{X})[k, (t_j)_{j\in\mathbf{k}}] \\
& {\scriptstyle\psi}\searrow & \downarrow {\scriptstyle\psi_2\mathring{\Box}\mathrm{id}} \\
& & (\mathcal{X}\mathring{\Box}\mathcal{X}\mathring{\Box}\mathcal{X})[n, (k_i)_{i\in\mathbf{n}}, (t_j)_{j\in\mathbf{k}}].
\end{array}
$$

such that $\psi_2 \in \mathsf{coEnd}(\mathcal{X})_2(n, (k_1, \ldots, k_n); k)$, i.e., $\psi_2 \colon \mathcal{X}[k] \to (\mathcal{X}\mathring{\Box}\mathcal{X})[n, (k_1, \ldots, k_n)]$ is Σ_k-equivariant. Said differently, there is an isomorphism

$$\mathsf{coEnd}(\mathcal{X})_3(\underline{p}; t) \cong \mathsf{coEnd}(\mathcal{X})_2(n, (k_i)_{i\in\mathbf{n}}; k) \otimes_{\Sigma_k} \mathsf{coEnd}(\mathcal{X})_2(k, (t_j)_{j\in\mathbf{k}}; t)$$

where Σ_k acts by shuffling the factors $t_1, \ldots, t_k$ of $\mathcal{X}_2(k, (t_j)_{j\in\mathbf{k}}; t)$ in accordance to the Σ_k equivariance of maps in $\mathcal{X}_2(n, (k_i)_{i\in\mathbf{n}}; k)$. In general, given a profile

$$\underline{p} = (n^1, (n_i^2)_{i\in\mathbf{n}^1}, \ldots, (n_i^\ell)_{i\in\mathbf{n}^{\ell-1}}) \in \mathbf{N}^{\hat{o}\ell}$$

the object $\mathsf{coEnd}(\mathcal{X})_\ell(\underline{p}; n^\ell)$ is isomorphic to

$$\mathsf{coEnd}(\mathcal{X})_2\big(n^1, (n_i^2)_{i\in\mathbf{n^1}}; n^2\big) \otimes_{\Sigma_{n^2}} \cdots \otimes_{\Sigma_{n^{\ell-1}}} \mathsf{coEnd}(\mathcal{X})_2\big(n^{\ell-1}, (n_i^\ell)_{i\in\mathbf{n^{\ell-1}}}; n^\ell\big). \qquad (5.4.13)$$

Remark 5.4.14. We would like to be able to say that $\mathsf{coEnd}(\mathcal{X})$ is a symmetric $\mathbf{N}_{\mathsf{lev}}$-operad for *any* cosimplicial symmetric sequence $\mathcal{X}$, however this seems to not be the case. The issue seems to be based on the potential non-invertibility of θ (and similarly how $\mathring{\square}$ fails to be a *strictly* monoidal product for cosimplicial symmetric sequences). However, there is a class of cosimplicial symmetric sequences on which we get the desired symmetric $\mathbf{N}_{\mathsf{lev}}$-structure on $\mathsf{coEnd}(\mathcal{X})$.

Let us say that $\mathcal{X}$ is Σ-*copowered* if there is a sequence $\{\mathcal{Y}[n]\}_{n\geq 0}$ of cosimplicial objects in C with

$$\mathcal{X}[n] = \Sigma_n{\cdot}\mathcal{Y}[n] \cong \mathbf{\Sigma}[n] \otimes \mathcal{Y}[n]$$

and such that the Σ_n action on $\mathcal{X}[n]$ is trivial on $\mathcal{Y}[n]$ for all n. In such case we write $\mathcal{X} = \Sigma{\cdot}\mathcal{Y}$. The benefit for us is that if $\mathcal{X}$ is Σ-copowered, then θ has an inverse (which is constructed in the following proposition), and so $\mathring{\square}$ *is* a monoidal product when restricted to Σ-copowered cosimplicial symmetric sequences.

Proposition 5.4.15. *If $\mathcal{X} \in \mathsf{SymSeq}_{\mathsf{C}}^{\Delta}$ is Σ-copowered, then $\mathsf{coEnd}(\mathcal{X})$ is a symmetric $\mathbf{N}_{\mathsf{lev}}$-operad.*

Proof. This argument is rather long and somewhat tedious, so we break it up into several steps. The first step is to show that $\mathsf{coEnd}(\mathcal{X})$ is symmetric, in fact Σ-copoweredness is not required for this part.

Let $\ell \geq 0$. The left action of $\mathcal{I}^\Sigma$ on $\mathsf{coEnd}(\mathcal{X})_\ell$ is obtained by Σ_t action on the maps $\mathcal{X}[t] \to \mathcal{X}^{\mathring{\square}\ell}[\underline{p}]$ which comprise $\mathsf{coEnd}(\mathcal{X})_\ell$. The right action of $\mathcal{I}^\Sigma \mathring{\circ} \cdots \mathring{\circ} \mathcal{I}^\Sigma$ on $\mathsf{coEnd}(\mathcal{X})_\ell$ is obtained, e.g., , at $\ell = 2$ as follows. For a profile $\underline{q} = (n, (k_1, \cdots, k_n))$, we observe

$$(\mathcal{I}^\Sigma \hat{\circ} \mathcal{I}^\Sigma)(\underline{q}; k) \cong \Sigma_n \ltimes (\Sigma_{k_1} \times \cdots \times \Sigma_{k_n}) \leq \Sigma_k$$

acts via the Σ_k-equivariance of

$$\mathcal{X}[k] \to \Sigma[k] \otimes_{H(k_1,\ldots,k_n)} \mathcal{X}[n] \square (\mathcal{X}[k_1] \otimes \cdots \otimes \mathcal{X}[k_n]).$$

The general case follows a similar argument.

Second, we produce a multiplication map

$$\xi \colon \mathsf{coEnd}(\mathcal{X}) \odot_\Sigma \mathsf{coEnd}(\mathcal{X}) \to \mathsf{coEnd}(\mathcal{X}).$$

Two ingredients are crucial to this step. First, is the existence of maps

$$\mu_{\ell_1,\ldots,\ell_k} \colon \mathcal{X}^{\dot\square \ell_1} \mathring{\square} \cdots \mathring{\square} \mathcal{X}^{\dot\square \ell_k} \to \mathcal{X}^{\dot\square \ell} \tag{5.4.16}$$

for each tuple $\ell_1, \ldots, \ell_k$ such that $\ell_1 + \cdots + \ell_k = \ell$ which are inverse to the induced map by θ (see (4.2.2)). It is this step for which Σ-copoweredness of $\mathcal{X}$ seems essential and such maps μ are granted by utilizing the structure of Oper. Write $\mathcal{X} = \Sigma \cdot \mathcal{Y}$ and for $\underline{p} = (n^1, \cdots, (n_i^k)_{i \in \mathbf{n}^{k-1}})$ set

$$\mathcal{Y}^{\square k}[\underline{p}] = \mathcal{Y}[n^1] \square \left(\bigotimes_{i \in \mathbf{n}^1} \mathcal{Y}[n_i^2] \right) \square \cdots \square \left(\bigotimes_{i \in \mathbf{n}^{k-1}} \mathcal{Y}[n_i^k] \right).$$

Note in the above, we are utilizing the box product for C^Δ which is strictly monoidal.

For simplicity we describe the map $\mu_{1,2} \colon \mathcal{X} \mathring{\square} (\mathcal{X} \mathring{\square} \mathcal{X}) \to \mathcal{X}^{\dot\square 3}$ and note the general case follows from a similar argument . Note that $\mathcal{X} \mathring{\square} (\mathcal{X} \mathring{\square} \mathcal{X})$ takes as inputs profiles of the form $(n, (\underline{p}_1, \cdots, \underline{p}_n))$ for some unordered list of profiles $\underline{p}_i \in \mathbf{N}^{\hat\square 2}$.

78

Fix a specific profile $(n, (\underline{p_1} \amalg \cdots \amalg \underline{p_n})) = \underline{p}$ and write $\underline{p_i} = (k_i, (t_{i,1}, \cdots, t_{i,k_i})) \in \mathbf{N}^{\hat{\circ}2}[t_i]$ for $i = 1, \ldots, n$. There is an inclusion induced as follows

$$\mathcal{X}\mathring{\square}(\mathcal{X}\mathring{\square}\mathcal{X}))[n, (\underline{p_1}, \cdots, \underline{p_n})] \tag{5.4.17}$$

$$\cong (\Sigma[n] \otimes \mathcal{Y}[n])\square \left((\Sigma^{\circ 2}[\underline{p_1}] \otimes \mathcal{Y}^{\square 2}[\underline{p_1}]) \otimes \cdots \otimes (\Sigma^{\circ 2}[\underline{p_n}] \otimes \mathcal{Y}^{\square 2}[\underline{p_n}]) \right)$$

$$\cong \left(\Sigma[n] \otimes_{\Sigma_n} \left(\coprod_{\dagger} \Sigma[t] \otimes_{\Sigma_{t_1} \times \cdots \times \Sigma_{t_n}} \Sigma^{\circ 2}[\underline{p_1}] \times \cdots \times \Sigma^{\circ 2}[\underline{p_n}] \right) \right) \otimes \mathcal{Y}^{\square 3}[\underline{p}]$$

$$\cong \Sigma^{\circ 3}[n, (\underline{p_1}, \cdots, \underline{p_n})] \otimes \mathcal{Y}^{\square 3}[\underline{p}] \xrightarrow{(*)} \Sigma^{\circ 3}[\underline{p}] \otimes \mathcal{Y}^{\square 3}[\underline{p}] \cong \mathcal{X}^{\mathring{\square}3}[\underline{p}]$$

where $\dagger$ runs over all Σ_n permutations of $t_1, \cdots, t_n$ and $(*)$ is induced by the natural inclusion $\iota \colon \Sigma^{\circ 3}[n, (\underline{p_1}, \cdots, \underline{p_n})] \to \Sigma^{\circ 3}[\underline{p}]$. Moreover, the map $\mu_{1,2}$ at profile $\underline{p}$ is then induced from the inclusion described above via the isomorphism

$$\left(\mathcal{X}\mathring{\square}(\mathcal{X}\mathring{\square}\mathcal{X}) \right)[\underline{p}] \cong \coprod_{(n, (\underline{p_1}\amalg\cdots\amalg\underline{p_n}))=\underline{p}} \left(\mathcal{X}\mathring{\square}(\mathcal{X}\mathring{\square}\mathcal{X}) \right)[n, (\underline{p_1}, \ldots, \underline{p_n})].$$

A straightforward computation then shows that $\mu_{1,2}$ is inverse to θ.

The second ingredient to producing ξ is a map

$$\mathsf{coEnd}(\mathcal{X})_{\ell_1}\hat{\circ}\cdots\hat{\circ}\,\mathsf{coEnd}(\mathcal{X})_{\ell_k} \xrightarrow{\Gamma} \mathrm{Map}_{\Delta^{\mathrm{res}}}\left(\mathcal{X}^{\mathring{\square}k}, \mathcal{X}^{\mathring{\square}\ell_1}\mathring{\square}\cdots\mathring{\square}\mathcal{X}^{\mathring{\square}\ell_k} \right)^{\Sigma} \tag{5.4.18}$$

which we construct as follows. Let $\alpha_i \colon \mathcal{X} \to \mathcal{X}^{\mathring{\square}\ell_i}$ for $i = 1, \ldots, k$. The map Γ is induced by the assignment $(\alpha_1, \ldots, \alpha_k) \mapsto \alpha_1\mathring{\square}\cdots\mathring{\square}\alpha_k$, where, e.g., if $k = 2$ and $\underline{p} = (n, (t_1, \cdots, t_n)) \in \mathbf{N}^{\hat{\circ}2}[t]$ then

$$(\alpha_1\mathring{\square}\alpha_2)[\underline{p}] \colon \mathcal{X}^{\mathring{\square}2}[\underline{p}] \to (\mathcal{X}^{\ell_1}\mathring{\square}\mathcal{X}^{\ell_2})[\underline{p}]$$

is obtained levelwise by the maps $\alpha_1[n] \colon \mathcal{X}[n] \to \mathcal{X}^{\mathring{\square}\ell_1}[n]$ and $\alpha_2[t_i] \colon \mathcal{X}^{\mathring{\square}\ell_2}[t_i]$ for $i = 1, \ldots, n$.

With these two ingredients in place, the composition ξ is obtained via the composition

$$\mathrm{Map}_{\Delta^{\mathrm{res}}}\left(\mathcal{X}, \mathcal{X}^{\dot{\square}k}\right)^{\Sigma} \dot{\otimes}_{\Sigma} \left(\mathrm{Map}_{\Delta^{\mathrm{res}}}\left(\mathcal{X}, \mathcal{X}^{\dot{\square}\ell_1}\right)^{\Sigma} \hat{\circ} \cdots \hat{\circ} \, \mathrm{Map}_{\Delta^{\mathrm{res}}}\left(\mathcal{X}, \mathcal{X}^{\dot{\square}\ell_k}\right)^{\Sigma}\right)$$

$$\xrightarrow{\mathrm{id}\dot{\otimes}_{\Sigma}\Gamma} \mathrm{Map}_{\Delta^{\mathrm{res}}}\left(\mathcal{X}, \mathcal{X}^{\dot{\square}k}\right)^{\Sigma} \dot{\otimes}_{\Sigma} \mathrm{Map}_{\Delta^{\mathrm{res}}}\left(\mathcal{X}^{\dot{\square}k}, \mathcal{X}^{\dot{\square}\ell_1}\dot{\square} \cdots \dot{\square}\mathcal{X}^{\dot{\square}\ell_k}\right)^{\Sigma}$$

$$\xrightarrow{\mathrm{comp.}} \mathrm{Map}_{\Delta^{\mathrm{res}}}\left(\mathcal{X}, \mathcal{X}^{\dot{\square}\ell_1}\dot{\square} \cdots \dot{\square}\mathcal{X}^{\dot{\square}\ell_k}\right)^{\Sigma}$$

$$\xrightarrow{(\mu_{\ell_1,\ldots,\ell_k})_*} \mathrm{Map}_{\Delta^{\mathrm{res}}}\left(\mathcal{X}, \mathcal{X}^{\dot{\square}\ell}\right)^{\Sigma}.$$

Fortunately, the unit map is simpler to describe. We obtain $\varepsilon\colon \mathcal{I}^{\Sigma} \to \mathsf{coEnd}(\mathcal{X})$ as the morphism

$$\Sigma[n] \to \mathrm{Map}_{\Delta^{\mathrm{res}}}(\mathcal{X}[n], \mathcal{X}[n])^{\Sigma_n}$$

adjoint to the action map $\Sigma[n] \otimes \mathcal{X}[n] \to \mathcal{X}[n]$ which expresses the Σ_n equivariance of $\mathcal{X}[n]$.

Showing that ξ and ε satisfy the appropriate associativity and unitality conditions is a tedious though ultimately straightforward and may be adapted from the (somewhat simpler) proof of Proposition 4.2.8 found in Section 6.1.1. $\qquad\square$

5.4.19   Oper-algebras are operads

Our aim is now to show that Oper-algebras indeed model (one-color) operads.

Proposition 5.4.20. *There is an equivalence of categories between algebras over* $\mathsf{Oper}^{\mathsf{C}}$ *and operads in* C.

Proof. We show that a symmetric Oper-algebra is necessarily an operad and note that the argument is readily reversed to show the converse statement. Suppose $\mathcal{W}$ is a symmetric Oper-algebra. Note, $\mathsf{Oper}_2\dot{\otimes}_{\Sigma}\mathcal{W}^{\dot{\circ}2} \to \mathcal{W}$ consists of maps

$$\mathsf{Oper}_2(n, (k_1, \ldots, k_n); k)\dot{\otimes}_{\Sigma} (\mathcal{W}[n] \otimes \mathcal{W}[k_1] \otimes \cdots \otimes \mathcal{W}[k_n]) \to \mathcal{W}[k] \tag{5.4.21}$$

for each $\underline{p} = (n, (k_1, \ldots, k_n)) \in \mathbf{N}^{\hat{\delta}2}$. Fix such a profile $\underline{p}$ and let $p_1, \ldots, p_m$ be the multiplicities of the distinct factors $d_1, \ldots, d_m$ among $k_1, \ldots, k_n$. Coequalizing the actions of $\mathcal{I}^{\Sigma}$ identifies the symmetric group actions (resp. with k_i replacing n)

$$\Sigma[n] \otimes \mathcal{W}[n] \to \mathcal{W}[n]$$

with the right action of $\mathcal{I}^{\Sigma}$ given in the proof of Proposition 5.4.15. Thus, (5.4.21) yields Σ_k-equivariant map of the form

$$\Sigma[k] \otimes_{H(k_1, \ldots, k_n)} \mathcal{W}[n] \otimes \mathcal{W}[k_1] \otimes \cdots \otimes \mathcal{W}[k_n] \to \mathcal{W}[k] \tag{5.4.22}$$

which moreover obeys the correct equivariance, e.g., as described in May [61]. Said differently, (5.4.22) is the factor $(\mathcal{W} \circ \mathcal{W})[n, (k_1, \ldots, k_n)]$ (as in Definition 2.1.6) and the collection of all such maps then pieces together to form

$$m \colon \mathcal{W} \circ \mathcal{W} \to \mathcal{W}.$$

Since $\mathcal{W} \in \mathsf{Alg}_{\mathsf{Oper}}$ there is a commutative diagram of the form

$$
\begin{array}{ccc}
(\mathsf{Oper}_2 \dot{\otimes}_{\Sigma}(\mathsf{Oper}_1 \hat{\circ} \mathsf{Oper}_2)) \dot{\otimes}_{\Sigma}(\mathcal{W}^{\hat{\delta}3}) & \xrightarrow{\cong} & \mathsf{Oper}_2 \dot{\otimes}_{\Sigma}((\mathsf{Oper}_1 \dot{\otimes}_{\Sigma}(\mathcal{W})) \hat{\circ} (\mathsf{Oper}_2 \dot{\otimes}_{\Sigma}(\mathcal{W}^{\hat{\delta}2}))) \\
\Big\downarrow{\scriptstyle \xi_{2,(1,2)} \dot{\otimes}_{\Sigma} \mathrm{id}} & & \Big\downarrow{\scriptstyle \mathrm{id} \dot{\otimes}_{\Sigma}(\mu_1 \hat{\circ} \mu_2)} \\
& & \mathsf{Oper}_2 \dot{\otimes}_{\Sigma}(\mathcal{W} \hat{\circ} \mathcal{W}) \\
& & \Big\downarrow{\scriptstyle \mu_2} \\
\mathsf{Oper}_3 \dot{\otimes}_{\Sigma}(\mathcal{W}^{\hat{\delta}3}) & \xrightarrow{\mu_3} & \mathcal{W}.
\end{array}
$$

The composite of the right side maps describes

$$\mathcal{W} \circ (\mathcal{W} \circ \mathcal{W}) \xrightarrow{\mathrm{id} \circ m} \mathcal{W} \circ \mathcal{W} \xrightarrow{m} \mathcal{W}$$

and by construction the bottom map describes

$$(\mathcal{W} \circ \mathcal{W}) \circ \mathcal{W} \xrightarrow{m \circ \mathrm{id}} \mathcal{W} \circ \mathcal{W} \xrightarrow{m} \mathcal{W}.$$

Associativity of m follows as $\xi_{2,(1,2)}$ is an isomorphism.

To produce the unit $u \colon I \to \mathcal{W}$ we first recall that

$$\mu_0 \colon I \cong \mathcal{P}_0 \dot{\otimes}_\Sigma (\mathcal{W}^{\hat{\circ} 0}) \to \mathcal{W}$$

provides the unit map u on $\mathcal{W}$. There is then a commuting diagram

$$
\begin{array}{ccc}
\mathsf{Oper}_2 \dot{\otimes}_\Sigma \left((\mathsf{Oper}_0 \dot{\otimes}_\Sigma (\mathcal{W}^{\hat{\circ}0})) \hat{\circ} (\mathsf{Oper}_1 \dot{\otimes}_\Sigma (\mathcal{W})) \right) & \xrightarrow{\mathrm{id} \dot{\otimes}_\Sigma (\mu_0 \hat{\circ} \mu_1)} & \mathsf{Oper}_2 \dot{\otimes}_\Sigma (\mathcal{W} \hat{\circ} \mathcal{W}) \\
\Big\downarrow{\scriptstyle \cong} & & \Big\downarrow{\scriptstyle \mu_2} \\
\left(\mathsf{Oper}_2 \dot{\otimes}_\Sigma (\mathsf{Oper}_0 \hat{\circ} \mathsf{Oper}_1) \right) \dot{\otimes}_\Sigma (\mathcal{W}) & & \\
\Big\downarrow{\scriptstyle \xi_1 \dot{\otimes}_\Sigma \mathrm{id}} & & \\
\mathsf{Oper}_1 \dot{\otimes}_\Sigma (\mathcal{W}) & \xrightarrow{\quad \mu_1 \quad} & \mathcal{W}
\end{array}
$$

the composite of top and right arrows of which results in

$$I \circ \mathcal{W} \xrightarrow{u \circ \mathrm{id}} \mathcal{W} \circ \mathcal{W} \xrightarrow{m} \mathcal{W}$$

and the left and bottom arrows are all isomorphisms. Commutativity of the other unitality diagram follows a similar analysis. $\qquad\square$

Corollary 5.4.23. *Let $\mathcal{W}$ be an operad, i.e., Oper-algebra. Let $M \in \mathsf{C}$ and denote by $\bar{M}$ the symmetric sequence concentrated at level 0 with $\bar{M}[0] = M$. Then, M is an $\mathcal{W}$-algebra (in the sense of Definition 5.3.11) if and only if M is an $\mathcal{W}$-algebra in the classic sense.*

Proof. As in Definition 5.3.11, a $\mathcal{W}$-algebra consists of maps

$$\mathsf{Oper}_\ell \dot\otimes_\Sigma (\mathcal{W}^{\hat{\circ}(\ell-1)} \hat\circ \bar{M}) \to \bar{M}.$$

Note, since $\bar{M}$ is concentrated at 0, the only nontrivial contributors to such maps will have profiles which end in a string of 0. In particular, for $\ell = 2$ there are maps of the form

$$\mathsf{Oper}_2(n, (0, \dots, 0); 0) \dot\otimes_\Sigma \left(\mathcal{W}[n] \otimes M^{\otimes n} \right) \to M.$$

Since $\mathsf{Oper}_2(n, (0, \dots, 0); 0) \cong \boldsymbol{\Sigma}[0] \cong \mathbf{1}$, the above maps descends to

$$\mathcal{W}[n] \otimes_{\Sigma_n} M^{\otimes n} \to M$$

after coequalizing. Associativity and unitality follow a similar argument as the proof of Proposition 5.4.20. $\qquad\square$

Remark 5.4.24. Though our description of Oper is new, descriptions of an $\mathbf{N}$-colored operad whose algebras are operads is not new. Berger-Moerdijk describe an $\mathbf{N}$-colored operad $\mathcal{M}_{\mathsf{Op}}$ in terms of trees whose algebras are operads in [13, 1.5.6] (see also Dehling-Vallette [32]). Applying the forgetful functor $\mathbb{U}$ from Section 5.2.24 to Oper yields an isomorphic $\mathbf{N}$-colored operad to that of Berger-Moerdijk, i.e., $\mathbb{U}\mathsf{Oper} \cong \mathcal{M}_{\mathsf{Op}}$.

5.4.25   A model for A_∞-operads

We will now focus on a particular coendomorphism $\mathbf{N}_{\mathsf{lev}}$-operad in Top, namely that on the cosimplicial symmetric sequence $\Sigma \cdot \Delta^\bullet$ with $\Sigma \cdot \Delta^\bullet[n] = \Sigma_n \cdot \Delta^\bullet$.

Proposition 5.4.26. *There is an equivalence of $\mathbf{N}_{\mathsf{lev}}$-operads* $\mathsf{coEnd}(\Sigma \cdot \Delta^\bullet) \to \mathsf{Oper}^{\mathsf{Top}}$.

Proof. Note that equivalences of $\mathbf{N}_{\mathsf{lev}}$-operads are computed levelwise (Definition 5.3.15) and that a morphism $f \colon X \to Y$ of cosimplicial objects in Top induces a map $(\Sigma \cdot X)^{\hat\square k} \to$

$(\Sigma \cdot Y)^{\mathring{\square}k}$ for $k \geq 1$. If additionally there is a retract $r \colon Y \to X$ of f there is a map $\mathsf{coEnd}(\Sigma \cdot X) \to \mathsf{coEnd}(\Sigma \cdot Y)$ on coendomorphism operads induced by post-composition with f and pre-composition with r.

Since there are morphisms $* \xrightarrow{\sim} \Delta^n \xrightarrow{\sim} *$ for all $n \geq 0$ (i.e., by inclusion at a vertex) we then have

$$\mathrm{Map}_{\mathbf{\Delta}^{\mathsf{res}}}\left(\Sigma \cdot \Delta^\bullet, (\Sigma \cdot \Delta^\bullet)^{\mathring{\square}k}\right)^\Sigma \xrightarrow{\;(\dagger)\;} \mathrm{Map}_{\mathbf{\Delta}^{\mathsf{res}}}\left(\Sigma \cdot \underline{*}, (\Sigma \cdot \underline{*})^{\mathring{\square}k}\right)^\Sigma \cong \mathrm{Map}\left(\Sigma, \Sigma^{\circ k}\right)^\Sigma$$

for all $k \geq 0$, where $\underline{*}$ denotes the constant cosimplicial object on $* \in \mathsf{Top}$. Moreover, since $* \to \Delta^n \to *$ consists of weak equivalences between fibrant and cofibrant objects for all n, the indicated map $(\dagger)$ is a weak equivalence in Top. $\qquad\square$

Note that for $\underline{p} \in \mathbf{N}^{\mathring{\circ}k}[t]$, $\mathsf{Oper}_k^{\mathsf{Top}}(\underline{p}; t)$ is just the discrete space $\Sigma^{\circ k}[\underline{p}]$. Similarly, $\mathsf{Oper}^{\mathsf{Top}_*} \cong \mathsf{Oper}_+^{\mathsf{Top}}$ will encode operads in $(\mathsf{Top}_*, \wedge, S^0)$ and thus also in Spt via the tensoring of Spt over Top_*.

Remark 5.4.27. Note the functor $(-)_+ \colon (\mathsf{Top}, \times, *) \to (\mathsf{Top}_*, \wedge, S^0)$ which adds a disjoint basepoint induces isomorphisms of pointed spaces

$$\mathrm{Map}_{\mathbf{\Delta}^{\mathsf{res}}}^{\mathsf{Top}_*}\left(\Sigma \cdot \Delta_+^\bullet, (\Sigma \cdot \Delta_+^\bullet)^{\mathring{\square}k}\right)^\Sigma \cong \mathrm{Map}_{\mathbf{\Delta}^{\mathsf{res}}}^{\mathsf{Top}}\left(\Sigma \cdot \Delta^\bullet, (\Sigma \cdot \Delta^\bullet)^{\mathring{\square}k}\right)_+^\Sigma.$$

Thus, there is an isomorphism

$$\mathsf{coEnd}(\Sigma \cdot \Delta_+^\bullet) \cong \mathsf{coEnd}(\Sigma \cdot \Delta^\bullet)_+$$

of $\mathbf{N}_{\mathsf{lev}}$-operads in Top_*. For ease of notation we write $\mathcal{A}$ for this $\mathbf{N}_{\mathsf{lev}}$-operad and note that Proposition 5.4.26 provides a map $\rho \colon \mathcal{A} \xrightarrow{\sim} \mathsf{Oper}^{\mathsf{Top}_*}$, i.e., $\mathcal{A}$ is a suitably "fattened-up" version of Oper which will encode A_∞-operads as its algebras, similar to $\mathbb{A}$ encoding A_∞-monoids in Example 4.1.6.

Chapter 6

Homotopy coherent operad structures on the derivatives of the identity

6.1 An operad structure for the derivatives of the identity in $\mathsf{Alg}_{\mathcal{O}}$

The aim of this final section is to prove Theorem 1.1.1. We begin by proving Proposition 4.2.8 which as a corollary provides a proof of the Theorem 1.1.1(a). In Section 6.1.3 we prove Theorem 1.1.1(b).

6.1.1 Proof of Theorem 1.1.1(a)

Since $C(\mathcal{O})$ is a $\mathring{\square}$-monoid (see Proposition 4.2.7), Theorem 1.1.1(a) will follow from Proposition 4.2.8, which we prove below.

Proof of Proposition 4.2.8. Let $\mathcal{X}$ be a $\mathring{\square}$-monoid in $\mathsf{SymSeq}_{\mathsf{Spt}}^{\triangle}$ whose multiplication we denote by $m\colon \mathcal{X}\mathring{\square}\mathcal{X} \to \mathcal{X}$. We aim to show that $\mathsf{Tot}\,\mathcal{X}$ is an algebra over $\mathcal{A}$. We define maps λ_ℓ as follows (note the notation $\dot{\wedge}$ as $\dot{\otimes}$ from Definition 5.2.14 for the monoidal category

85

$(\mathsf{Spt}, \wedge, S))$

$$\lambda_\ell \colon \mathcal{A}_\ell \dot{\wedge}_\Sigma (\operatorname{Tot} \mathcal{X})^{\hat{\circ}\ell} \to \operatorname{Tot} \mathcal{X}$$

For simplicity we first describe the $\ell = 2$ case. Let $\underline{p} = (n, (k_1, \dots, k_n)) \in \mathbf{N}^{\hat{\circ}2}[k]$. Let $\psi \in \mathcal{A}_2(\underline{p}; t)$, and let $\alpha, \beta \colon \Sigma{\cdot}\Delta_+^\bullet \to \mathcal{X}$ be maps of cosimplicial symmetric sequences. Define γ at level k by the composite

$$
\begin{array}{ccc}
(\Sigma{\cdot}\Delta_+^\bullet)^{\hat{\Box}2}[n, (k_1, \dots, k_n)] & \xrightarrow{\alpha[n]\hat{\Box}\beta[k_1,\dots,k_n]} & (\mathcal{X}^{\hat{\Box}2})[n, (k_1, \dots, k_n)] \\
\psi[k] \big\uparrow & & \big\downarrow m_* \\
(\Sigma{\cdot}\Delta_+^\bullet)[k] & \xrightarrow{\quad\gamma[k]\quad} & \mathcal{X}[k]
\end{array}
$$

where $\alpha[n]\mathring{\Box}\beta[k_1, \dots, k_n]$ is provided via the map Γ from $(5.4.18)$, the construction of which may be readily altered to give a map

$$\Gamma \colon \left(\operatorname{Map}^{\mathsf{Spt}}_{\mathbf{\Delta}^{\mathsf{res}}} (\Sigma{\cdot}\Delta_+^\bullet, \mathcal{X})^\Sigma \right)^{\hat{\circ}\ell} \to \operatorname{Map}^{\mathsf{Spt}}_{\mathbf{\Delta}^{\mathsf{res}}} \left((\Sigma{\cdot}\Delta_+^\bullet)^{\hat{\Box}\ell}, \mathcal{X}^{\hat{\Box}\ell} \right)^\Sigma.$$

In general, λ_ℓ is given by the following composite (compare with $[3, (1.13)]$)

$$\operatorname{Map}^{\mathsf{Top}_*}_{\mathbf{\Delta}^{\mathsf{res}}} \left(\Sigma{\cdot}\Delta_+^\bullet, (\Sigma{\cdot}\Delta_+^\bullet)^{\hat{\Box}\ell} \right)^\Sigma \dot{\wedge}_\Sigma \left(\operatorname{Map}^{\mathsf{Spt}}_{\mathbf{\Delta}^{\mathsf{res}}} (\Sigma{\cdot}\Delta_+^\bullet, \mathcal{X})^\Sigma \right)^{\hat{\circ}\ell}$$

$$\xrightarrow{\operatorname{id}\wedge_\Sigma\Gamma} \operatorname{Map}^{\mathsf{Top}_*}_{\mathbf{\Delta}^{\mathsf{res}}} \left(\Sigma{\cdot}\Delta_+^\bullet, (\Sigma{\cdot}\Delta_+^\bullet)^{\hat{\Box}\ell} \right)^\Sigma \dot{\wedge}_\Sigma \operatorname{Map}^{\mathsf{Spt}}_{\mathbf{\Delta}^{\mathsf{res}}} \left((\Sigma{\cdot}\Delta_+^\bullet)^{\hat{\Box}\ell}, \mathcal{X}^{\hat{\Box}\ell} \right)^\Sigma$$

$$\xrightarrow{\operatorname{compose}} \operatorname{Map}^{\mathsf{Spt}}_{\mathbf{\Delta}^{\mathsf{res}}} \left(\Sigma{\cdot}\Delta_+^\bullet, \mathcal{X}^{\hat{\Box}\ell} \right)^\Sigma$$

$$\xrightarrow{m_*} \operatorname{Map}^{\mathsf{Spt}}_{\mathbf{\Delta}^{\mathsf{res}}} \left(\Sigma{\cdot}\Delta_+^\bullet, \mathcal{X} \right)^\Sigma$$

where the composition map is adjoint to the composite of evaluation maps

$$\Sigma{\cdot}\Delta_+^\bullet \wedge \operatorname{Map}^{\mathsf{Top}_*}_{\mathbf{\Delta}^{\mathsf{res}}}(\Sigma{\cdot}\Delta_+^\bullet, (\Sigma{\cdot}\Delta_+^\bullet)^{\hat{\Box}\ell})^\Sigma \to (\Sigma{\cdot}\Delta^\bullet)^{\hat{\Box}\ell}_+, \tag{6.1.2}$$

$$(\Sigma{\cdot}\Delta_+^\bullet)^{\hat{\Box}\ell} \dot{\wedge}_\Sigma \operatorname{Map}^{\mathsf{Spt}}_{\mathbf{\Delta}^{\mathsf{res}}}((\Sigma{\cdot}\Delta_+^\bullet)^{\hat{\Box}\ell}, \mathcal{X}^{\hat{\Box}\ell})^\Sigma \to \mathcal{X}^{\hat{\Box}\ell}.$$

and m_* is induced by the $\dot{\square}$-monoid structure on $\mathcal{X}$.

To show that λ is associative we consider the following diagram, with $\psi' \in \mathcal{A}_n$, $\psi_i \in \mathcal{A}_{k_i}$ for $i = 1, \ldots, n$ such that the composite $\xi(\psi'; \psi_1, \ldots, \psi_n) = \psi \in \mathcal{A}_k$.

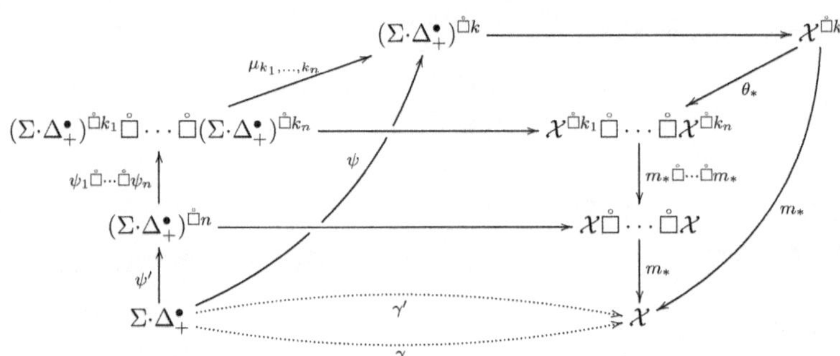

Note here that m_* is induced by repeatedly applying the pairing $m \colon \mathcal{X}\dot{\square}\mathcal{X} \to \mathcal{X}$ from the left, i.e.,

$$\mathcal{X}\dot{\square}\mathcal{X}\dot{\square}\cdots\dot{\square}\mathcal{X} \xrightarrow{m\dot{\square}\mathrm{id}\dot{\square}\cdots\dot{\square}\mathrm{id}} \cdots \xrightarrow{m\dot{\square}\mathrm{id}\dot{\square}\mathrm{id}} \mathcal{X}\dot{\square}\mathcal{X}\dot{\square}\mathcal{X} \xrightarrow{m\dot{\square}\mathrm{id}} \mathcal{X}\dot{\square}\mathcal{X} \xrightarrow{m} \mathcal{X}.$$

The dashed morphisms γ and γ' are induced by $\lambda_k \xi_{n,(k_1,\ldots,k_n)}$ and $\lambda_n(\mathrm{id}\dot{\otimes}_\Sigma(\lambda_{k_1}\hat{\circ}\cdots\hat{\circ}\lambda_{k_n}))$, respectively. Note as well that $\mu_{k_1,\ldots,k_n}$ is as in the proof of Proposition 5.4.15, and θ_* is the grouping map induced by θ (see Section 4.2.3), by which it follows that γ and γ' must agree.

For unitality we recall that $\epsilon \colon \mathcal{I}^\Sigma \to \mathcal{A}$ is induced by the inclusion at id_Δ and therefore the composite $\lambda_1[n]\epsilon[n]$ in the following diagram

$$\mathcal{A}_1(n;n) \wedge_{\Sigma_n} \mathrm{Map}^{\mathsf{Spt}}_{\Delta^{\mathrm{res}}}\left(\Sigma{\cdot}\Delta^\bullet_+[n], \mathcal{X}[n]\right)^{\Sigma_n} \xrightarrow{\lambda_1[n]} \mathrm{Map}^{\mathsf{Spt}}_{\Delta^{\mathrm{res}}}\left(\Sigma{\cdot}\Delta^\bullet_+[n], \mathcal{X}[n]\right)^{\Sigma_n}$$

$$\epsilon[n]\Big\uparrow \qquad\qquad \cong \qquad\qquad$$

$$(\Sigma_n)_+ \wedge_{\Sigma_n} \mathrm{Map}^{\mathsf{Spt}}_{\Delta^{\mathrm{res}}}\left(\Sigma{\cdot}\Delta^\bullet_+[n], \mathcal{X}[n]\right)^{\Sigma_n}$$

87

is given by $S^0 \wedge \mathrm{Tot}\,\mathcal{X}[n] \xrightarrow{\cong} \mathrm{Tot}\,\mathcal{X}[n]$. $\qquad\qquad\qquad\qquad\qquad\qquad\qquad\square$

6.1.3 An equivalence of A_∞-operads between $\mathcal{O}$ and $\partial_* \mathrm{Id}_{\mathsf{Alg}_\mathcal{O}}$

We now show that the induced operad structure on $\partial_* \mathrm{Id}_{\mathsf{Alg}_\mathcal{O}}$ from Proposition 4.2.8 agrees with the induced $\mathcal{A}$-algebra structure on $\mathcal{O}$, thus proving Theorem 1.1.1(b). Let $\rho \colon \mathcal{A} \xrightarrow{\sim}$ Oper be the map described in Remark 5.4.27 and note an operad $\mathcal{O} \in \mathsf{Alg}_{\mathsf{Oper}}$ is in algebra over $\mathcal{A}$ via the forgetful functor ρ^*.

Proof of Theorem 1.1.1(b). By equivalence of A_∞-operads we mean equivalence of $\mathcal{A}$-algebras which restricts to an equivalence of underlying symmetric sequences.

Recall there is a natural coaugmentation $\mathcal{O} \to C(\mathcal{O})$ via $\mathcal{O} \to J$. We have shown in Section 3.3.5 that the coface k-cubes associated to

$$\mathcal{O} \to C(\mathcal{O}) \quad \text{and} \quad \partial_* \mathrm{Id}_{\mathsf{Alg}_\mathcal{O}} \to \mathrm{holim}_{\mathbf{\Delta}^{\leq n-1}} \partial_*((UQ)^{\bullet+1})$$

are equivalent. Denoting these k-cubes by $\mathcal{X}_k$ and $\mathcal{Y}_k$, respectively, we note for $k \geq n \geq 1$ that as $\mathcal{Y}_k[n]$ is homotopy cartesian so is $\mathcal{X}_k[n]$. That is to say, for all $n \geq 1$

$$\mathcal{O}[n] \xrightarrow{\sim} \mathrm{holim}_{\mathbf{\Delta}}(C(\mathcal{O})[n]).$$

Let $\underline{\mathcal{O}}$ be the constant cosimplicial object in SymSeq on $\mathcal{O}$. From the above, the coaugmentation $\mathcal{O} \to C(\mathcal{O})$ induces a map of cosimplicial symmetric sequences $\varphi \colon \underline{\mathcal{O}} \to C(\mathcal{O})$ such that $\mathrm{Tot}\,\underline{\mathcal{O}} \xrightarrow{\sim} \mathrm{Tot}\,C(\mathcal{O})$. Moreover, $\underline{\mathcal{O}}$ inherits a natural $\mathring{\square}$-monoid structure induced by the operad structure maps $\mathcal{O} \circ \mathcal{O} \to \mathcal{O}$ and $I \to \mathcal{O}$, and φ respects this structure (i.e., is a map of $\mathring{\square}$-monoids).

For each $n \geq 0$ we have

$$\mathrm{Map}^{\mathsf{Spt}}_{\boldsymbol{\Delta}^{\mathsf{res}}}(\Sigma_n{\cdot}\boldsymbol{\Delta}^{\bullet}_+, \underline{\mathcal{O}}[n])^{\Sigma_n} \xrightarrow{\sim} \mathrm{Map}^{\mathsf{Spt}}_{\boldsymbol{\Delta}^{\mathsf{res}}}(\Sigma_n{\cdot}\underline{\boldsymbol{\Delta}}^0_+, \underline{\mathcal{O}}[n])^{\Sigma_n}$$

$$\cong \mathrm{Map}^{\mathsf{Spt}}(\Sigma_n{\cdot}S^0, \mathcal{O}[n])^{\Sigma_n} \cong \mathrm{Map}^{\mathsf{Spt}}(S^0, \mathcal{O}[n]) \cong \mathcal{O}[n].$$

and therefore, $\mathsf{Tot}\,\underline{\mathcal{O}} \xrightarrow{\sim} \mathcal{O}$. Thus, there are commuting diagrams for all $n \geq 0$

$$\begin{array}{ccccc}
\mathcal{A}_n \dot{\otimes}_\Sigma (\rho^*\mathcal{O})^{\hat{\otimes}n} & \longleftarrow & \mathcal{A}_n \dot{\otimes}_\Sigma (\mathsf{Tot}\,\underline{\mathcal{O}})^{\hat{\otimes}n} & \longrightarrow & \mathcal{A}_n \dot{\otimes}_\Sigma (\mathsf{Tot}\,C(\mathcal{O}))^{\hat{\otimes}n} \\
\downarrow & & \downarrow & & \downarrow \\
\rho^*\mathcal{O} & \xleftarrow{\quad\sim\quad} & \mathsf{Tot}\,\underline{\mathcal{O}} & \xrightarrow{\quad\sim\quad} & \mathsf{Tot}\,C(\mathcal{O})
\end{array} \qquad (6.1.4)$$

where the left is the $\mathcal{A}$-algebra structure map on $\rho^*\mathcal{O}$ (which must factor through Oper) and the right is the $\mathcal{A}$-algebra structure map on $\partial_*\mathtt{Id}_{\mathsf{Alg}_\mathcal{O}}$. $\qquad\square$

6.1.5   A class of $\partial_*\mathtt{Id}_{\mathsf{Alg}_\mathcal{O}}$-algebras

Though it will follow abstractly from Theorem 6.1.3, the following corollary show that it is possible to describe an action of $\partial_*\mathtt{Id}_{\mathsf{Alg}_\mathcal{O}}$ explicitly on the TQ-completion of sufficiently connected $\mathcal{O}$-algebras. Recall that $X \simeq X^{\wedge}_{\mathsf{TQ}}$ for 0-connected $X \in \mathsf{Alg}_\mathcal{O}$.

Corollary 6.1.6. *Any 0-connected $\mathcal{O}$-algebra X is weakly equivalent to an algebra over* $\partial_*\mathtt{Id}_{\mathsf{Alg}_\mathcal{O}}$ *via* $X \mapsto X^{\wedge}_{\mathsf{TQ}}$.

Proof. A straightforward modification of the proof of Proposition 4.2.7 permits a well-defined map of cosimplicial diagrams

$$r \colon C(\mathcal{O}) \overset{\circ}{\square} C(X) \to C(X)$$

which endows $C(X)$ with the structure of a left module over $C(\mathcal{O})$. Strictly speaking we do need to be careful here, as $C(\mathcal{O})$ is not a strict monoid, so the module structure is obtained by replacing the right-most instances of $C(\mathcal{O})$ with $C(X)$ in (4.2.4) and (4.2.5). Nonetheless, a straightforward adaptation of the proof of Proposition 4.2.8 demonstrates maps

$$\mathcal{A}_\ell \dot{\otimes}_\Sigma \left((\mathsf{Tot}\, C(\mathcal{O}))^{\hat{\delta}(\ell-1)} \hat{\circ} \bar{X}^\wedge_{\mathsf{TQ}} \right) \to \bar{X}^\wedge_{\mathsf{TQ}}$$

where $\bar{X}^\wedge_{\mathsf{TQ}}$ is the symmetric sequence concentrated at level 0 with value $X^\wedge_{\mathsf{TQ}}$, as required of Definition 5.3.11. $\qquad\square$

Remark 6.1.7. One intent of the above is to motivate the analogous statement for algebras over the derivatives of the identity in *spaces*, which *a priori* seems a bit more mysterious. Using the model $\partial_* \mathsf{Id}_{\mathsf{Top}_*} = \mathrm{holim}_\Delta\, C(\underline{S})$ we further show in the following section that for any $\underline{S}$-coalgebra Y in spectra (e.g., $Y = \Sigma^\infty X$) the derived primitives $\mathsf{Prim}_{\underline{S}}(Y)$ inherits the structure of an algebra over $\partial_* \mathsf{Id}_{\mathsf{Top}_*}$ via a pairing of cosimplicial objects with respect to $\overset{\circ}{\square}$ (see also [22], [50], [12]).

In this framework, Corollary 6.1.6 tells us that any 0-connected $X \in \mathsf{Alg}_\mathcal{O}$ is equivalent to its derived primitives $\mathsf{Prim}_{B(\mathcal{O})}(\mathsf{TQ}(X))$ (with respect to a suitable coalgebra structure on $B(\mathcal{O})$, see Section 3.3.4) as $\partial_* \mathsf{Id}_{\mathsf{Alg}_\mathcal{O}} \simeq \mathcal{O}$-algebras. Note also that $\mathsf{Prim}_{B(\mathcal{O})}(\mathsf{TQ}(X)) \simeq X^\wedge_{\mathsf{TQ}}$. As such, one possible future avenue for our work is to try to push this result to work for any nilpotent $\mathcal{O}$-algebra. This could potentially be used to prove that any nilpotent $\mathcal{O}$-algebra is equivalent to its TQ-completion (see also Section 6.3.6, along with [14], [68]; and further compare with [19], [6], [21], [16] that any nilpotent space is equivalent to its completion with respect to $\Omega^\infty\Sigma^\infty$).

6.2 An operad structure on the derivatives of the identity in spaces

The aim of this sectin is to show that the derivatives of the identity in spaces can be given an operad structure in a similar manner as with the operad structure just constructed for $\partial_* \mathrm{Id}_{\mathsf{Alg}_{\mathcal{O}}}$. As before, our method is to show that $C(\underline{S})$ admits $\mathring{\square}$-monoidal structure.

We need a technical lemma first; note the following is similar to the "tree ungrafting" argument from [22]. Recall that $P(n)$ denotes the underlying simplicial set of the n-th partition poset complex $\mathsf{Par}(n)$ (Definition 3.4.5).

Lemma 6.2.1. *For $k, p, q \geq 0$, there is a Σ_k-equivariant "decomposition" map*

$$\Psi_{k,(p,q)} \colon P(k)_{p+q} \to \prod_{\alpha \in P(k)_2} P(n)_p \times P(k_1)_q \times \cdots \times P(k_n)_q$$

where n and $k_1, \ldots, k_n$ are obtained by setting $|\alpha| = (n, (k_1, \ldots, k_n))$.

Proof. Let $n \geq 1$ and $T_1, \ldots, T_n$ be a partition of $\mathbf{k}$. Let $\beta_j \in P(T_j)_q$ be given by $\mu_0^j \leq \cdots \leq \mu_q^j$ for $j = 1, \ldots, n$, $\gamma \in P(n)_p$ given by $\lambda_0 \leq \cdots \leq \lambda_p$, and let λ_j' denote the partition obtained by replacing a set $\{\gamma_s\}_{s \in S} \in \lambda_j$ by $\coprod_{s \in S} T_s$. There is then an element $\gamma \circ (\beta_1, \cdots, \beta_n) \in P(k)_{p+q}$ given by

$$\lambda_0' \leq \cdots \leq \lambda_{p-1}' \leq \lambda_p' \cong \coprod_{i=1}^{n} \mu_0^i \leq \coprod_{i=1}^{n} \mu_1^i \leq \cdots \leq \coprod_{i=1}^{n} \mu_q^i.$$

Given $\alpha \in P(k)_2$, let $T_1, \cdots, T_n$ be the corresponding partition of $\mathbf{k}$ determined. Given $\gamma' \in P(k)_{p+q}$, the image $\Psi_{k,(p,q)}(\gamma')$ at the α-factor of the product is defined to be the string $(\gamma, \beta_1, \cdots, \beta_n)$ if there is a decomposition $\gamma' = \gamma \circ (\beta_1, \cdots, \beta_n)$ where $\gamma \in P(n)_p$ and $\beta_j \in P(T_j)_q$ for $j = 1, \ldots, n$, and by the basepoint otherwise. $\square$

Proposition 6.2.2. *For $p, q \geq 0$ there are maps $m_{p,q} \colon C(\underline{S})^p \circ C(\underline{S})^q \longrightarrow C(\underline{S})^{p+q}$.*

Proof. Let $p, q \geq 0$. For $k \geq 1$, $m_{p,q}$ at level k is given by the following composite

$$(C(\underline{S})^p \circ C(\underline{S})^q)[k] = \bigvee_{\alpha \in P(k)_2} \left(\left(\prod_{P(n)_p} S \right) \wedge \bigwedge_{i=1}^{n} \left(\prod_{P(k_i)_q} S \right) \right)$$

$$\to \bigvee_{\alpha \in P(k)_2} \left(\prod_{P(n)_p} S \wedge \prod_{P(k_1)_q \times \cdots \times P(k_n)_q} S \wedge \cdots \wedge S \right)$$

$$\to \prod_{\prod_{\alpha \in P(k)_2} P(n)_p \times P(k_1)_q \times \cdots \times P(k_n)_q} S \wedge S \wedge \cdots \wedge S$$

$$\xrightarrow{\Psi^*_{k,(p,q)}} \prod_{P(k)_{p+q}} S \wedge S \wedge \cdots \wedge S \cong C(\underline{S})^{p+q}[k]$$

where $n, k_1, \ldots, k_n$ are such that $|\alpha| = (n, (k_1, \ldots, k_n))$. $\qquad\square$

Proposition 6.2.3. *The cosimplicial symmetric sequence $C(\underline{S})$ admits a natural $\mathring{\Box}$-monoid structure, i.e., there are maps $m\colon C(\underline{S})\mathring{\Box}C(\underline{S}) \to C(\underline{S})$ and $u\colon \underline{I} \to C(\underline{S})$ which satisfy associativity and unitality.*

Proof. The map m is induced as follows. For $p + q = n$, the maps described in Proposition 6.2.2 fit into the following commuting squares

$$\begin{array}{ccc}
C(\underline{S})^p \circ C(\underline{S})^{q+1} & \xrightarrow{m_{p,q+1}} & C(\underline{S})^{p+q+1} \\
{\scriptstyle \mathrm{id}\circ d^0}\big\uparrow & & \big\uparrow{\scriptstyle m_{p+1,q}} \\
C(\underline{S})^p \circ C(\underline{S})^q & \xrightarrow{d^{p+1}\circ \mathrm{id}} & C(\underline{S})^{p+1} \circ C(\underline{S})^q
\end{array}$$

which induce m at level $n + 1$ upon taking colimits.

The unit map is induced by the coaugmentation $I \to C(\underline{S})$ given by the identity on I. Associativity and unitality of m and u follow from the same argument as in the proof of 4.2.7 with $C(\mathcal{O})^{12}$ replaced by $C(\underline{S})$. $\qquad\square$

---

[12]Though, perhaps to reflect the notation in this section, the cosimplicial object $C(\mathcal{O})$ should be written as $C(B(\mathcal{O}))$, as the bar construction $B(\mathcal{O})$ is a cooperad [22], *not* $\mathcal{O}$.

Proof of Theorem 1.1.4. From Proposition B.2.6, we know $\partial_* \mathrm{Id}_{\mathsf{Top}_*} \simeq \mathsf{Tot}\, C(\underline{S})$. The theorem then follows from Proposition 4.2.8. $\qquad\square$

6.2.4 Derived primitives are spectral Lie algebras

Arone-Ching have shown that the operad $\partial_* \mathrm{Id}_{\mathsf{Top}_*}$ plays a central role in homotopy theory [3], [4]. We now show explicitly that the derived primitives of a $\underline{S}$-coalgebra Y admit a left action by this operad. In particular, this defines a functor

$$\Xi \colon \mathsf{Top}_* \to \mathsf{Alg}_{\partial_* \mathrm{Id}}, \qquad X \mapsto \mathsf{Tot}\, C(\Sigma^\infty X)$$

which is closely related to work of Heuts as inducing equivalences after certain chromatic localizations [50].

Proof of Theorem 1.1.5. The idea is to show that $C(Y)$ is a left $\mathring{\Box}$-module over $C(\underline{S})$ concentrated in symmetric sequence level 0. That is, the following associativity (6.2.5) and unitality (6.2.6) diagrams commute

$$C(\underline{S})\mathbin{\mathring{\Box}}C(\underline{S})\mathbin{\mathring{\Box}}C(Y) \xrightarrow{\ \theta\ } C(\underline{S})\mathbin{\mathring{\Box}}(C(\underline{S})\mathbin{\mathring{\Box}}C(Y)) \xrightarrow{\ \mathrm{id}\mathbin{\mathring{\Box}}\mu\ } C(\underline{S})\mathbin{\mathring{\Box}}C(Y) \tag{6.2.5}$$

with left vertical map $\cong$, $(C(\underline{S})\mathbin{\mathring{\Box}}C(\underline{S}))\mathbin{\mathring{\Box}}C(Y) \xrightarrow{\ m\mathbin{\mathring{\Box}}\mathrm{id}\ } C(\underline{S})\mathbin{\mathring{\Box}}C(Y) \xrightarrow{\ \mu\ } C(Y)$, and right vertical map μ

and

$$C(\underline{S})\mathbin{\mathring{\Box}}C(Y) \xrightarrow{\ \mu\ } C(Y) \tag{6.2.6}$$

with $\underline{I}\mathbin{\mathring{\Box}}C(Y) \xrightarrow{\ u\mathbin{\mathring{\Box}}\mathrm{id}\ } C(\underline{S})\mathbin{\mathring{\Box}}C(Y)$ and $\underline{I}\mathbin{\mathring{\Box}}C(Y) \xrightarrow{\ \cong\ } C(Y)$

Once we have shown this structure, it follows that $\mathsf{Tot}\, C(Y)$ is a $\partial_* \mathrm{Id}_{\mathsf{Top}_*}$-algebra. We first produce maps $\mu_{p,q} \colon C(\underline{S})^p \circ C(Y)^q \to C(Y)^{p+q}$.

93

Let $p, q \geq 0$ be given, $\mu_{p,q}$ is then the composite

$$C(\underline{S})^p \circ C(Y)^q = \bigvee_{n \geq 1} C(\underline{S})^p[n] \wedge_{\Sigma_n} (C(Y)^q)^n$$

$$\to \prod_{n \geq 1} \left(\prod_{P(n)_p} S \right) \wedge_{\Sigma_n} \left(\bigwedge_{i=1}^{n} \prod_{k_i \geq 1} \left(\prod_{P(k_i)_q} Y^{\wedge k_i} \right)_{\Sigma_{k_i}} \right)$$

$$\to \prod_{n \geq 1} \left(\prod_{P(n)_p} S \right) \wedge_{\Sigma_n} \left(\prod_{k \geq 1} \prod_{k=k_1+\cdots+k_n} \bigwedge_{i=1}^{n} \left(\prod_{P(k_i)_q} Y^{k_i} \right)_{\Sigma_{k_i}} \right)$$

$$\to \prod_{k \geq 1} \prod_{n \geq 1} \left(\prod_{P(n)_p} S \right) \wedge_{\Sigma_n} \left(\prod_{k=k_1+\cdots+k_n} \left(\prod_{P(k_1)_q \times \cdots \times P(k_n)_q} \bigwedge_{i=1}^{n} Y^{\wedge k_i} \right)_{\Sigma_{k_1} \times \cdots \times \Sigma_{k_n}} \right)$$

$$\cong \prod_{k \geq 1} \left(\prod_{\alpha \in P(k)_2} \prod_{P(n)_p \times P(k_1)_q \times \cdots \times P(k_n)_q} Y^{\wedge k} \right)_{\Sigma_k}$$

$$\xrightarrow{\Psi^*_{k,(p,q)}} \prod_{k \geq 1} \left(\prod_{P(k)_{p+q}} Y^{\wedge k} \right)_{\Sigma_k}$$

The map $\mu \colon C(\underline{S}) \mathring{\square} C(Y) \to C(Y)$ is induced at cosimplicial level 0 by

$$C(\underline{S})^0 \circ C(Y)^0 \cong S \wedge Y \xrightarrow{\cong} Y$$

and for $n \geq 0$ at level $n+1$ by the commuting squares

$$\begin{array}{ccc}
C(\underline{S})^p \circ C(Y)^{q+1} & \xrightarrow{\mu_{p,q+1}} & C(Y)^{p+q+1} \\
{\scriptstyle \mathrm{id} \circ d^{q+1}} \uparrow & & \uparrow {\scriptstyle \mu_{p+1,q}} \\
C(\underline{S})^p \circ C(Y)^q & \xrightarrow{d^0 \circ \mathrm{id}} & C(\underline{S})^{p+1} \circ C(Y)^q
\end{array}$$

Here $p, q \geq 0$ are such that $p + q = n$.

Associativity and unitality of μ again follows from a similar straightforward modification of the argument from Proposition 4.2.7 (see also Corollary 6.1.6). $\qquad\square$

Remark 6.2.7. It is worth remarking that a straightforward modification to the proofs of Theorems 1.1.4 and 1.1.5 presented in this document shows that $\mathsf{Tot}\,C(\mathcal{Q})$ is an A_∞-operad and, if Y is a $\mathcal{Q}$-coalgebra, that $C(Y)$ is an algebra over this operad. In particular, our constructions provide a point-set model for $\mathsf{Tot}\,C(\mathcal{Q})$ as a (homotopy coherent) operad which is *Koszul dual* to $\mathcal{Q}$, and $Y \to \mathsf{Tot}\,C(Y)$ provides a comparison from $\mathcal{Q}$-coalgebras to algebras over this operad (see [38]).

6.3 Conjectures and future work

6.3.1 A chain rule for functors of structured ring spectra

We expect that our techniques from Section 6.1 which underlie our proof that the derivatives of the identity in $\mathsf{Alg}_{\mathcal{O}}$ form an operad should allow for a description of a "highly homotopy coherent chain rule" for functors of structured ring spectra. Essentially, a chain rule describes an equivalence of Goodwillie derivatives

$$\partial_* F \circ_{\partial_* \mathrm{Id}_\mathsf{D}} \partial_* G \simeq \partial_*(FG) \tag{6.3.2}$$

for functors $\mathsf{C} \xrightarrow{G} \mathsf{D} \xrightarrow{F} \mathsf{E}$ (where $\mathsf{C}, \mathsf{D}, \mathsf{E}$ are suitable model categories such as Spt or Top_*).

Arone-Ching [2] give a full description of a chain rule for functors between Top_* and Spt, building off a chain rule for functors of Spt described by Ching [23] (see also Klein-Rognes [53] and Yeakel [71]). The chain rule may be thought of a homotopical analog to the Faà di Bruno formula [37] for functions of a single real variable, which describes the n-th derivative of a composition $f \circ g$ in terms of the derivatives of f and g separately. Note that by setting either F or G to be the identity in (6.3.2), the chain rule describes a $(\partial_* \mathrm{Id}_\mathsf{D}, \partial_* \mathrm{Id}_\mathsf{C})$-bimodule structure on the derivatives of a homotopy functor $\mathsf{C} \to \mathsf{D}$[13].

Conjecture 6.3.3. *Let $\mathcal{O}, \mathcal{O}'$, and $\mathcal{O}''$ be operads of spectra and F, G be finitary simplicial functors* $\mathsf{Alg}_{\mathcal{O}} \xrightarrow{G} \mathsf{Alg}_{\mathcal{O}'} \xrightarrow{F} \mathsf{Alg}_{\mathcal{O}''}$. *Then there exists*

(i) A "highly homotopy coherent" chain rule map $\partial_* F \circ \partial_* G \to \partial_*(FG)$

(ii) An equivalence of $(\mathcal{O}'', \mathcal{O})$-bimodules of the form $\partial_* F \circ_{\mathcal{O}'} \partial_* G \simeq \partial_*(FG)$

Our method is to resolve the functors F and G by the stabilization adjunction, as done before with the identity, in order to build the desired map 6.3.3(i) above by a $\mathring{\square}$ pairing of cosimplicial symmetric sequences. When F and G are determined by a cofibrant $(\mathcal{O}'', \mathcal{O}')$-bimodule M and cofibrant $(\mathcal{O}', \mathcal{O})$-bimodule N, respectively, our techniques provide the desired result. Specifically, we require that $F \simeq |\operatorname{Bar}(M, \mathcal{O}, -)|$ and $G \simeq |\operatorname{Bar}(N, \mathcal{O}', -)|$ and that M, N consist of terms which are (-1)-connected. The desired map 6.3.3(i) then necessarily takes the form

$$\mathcal{A}_2 \dot{\otimes}_\Sigma (\partial_* F \hat{\circ} \partial_* G) \to \partial_*(FG).$$

In this case, however, since $\partial_* F \simeq M$, $\partial_* G \simeq N$, and $\partial_*(FG) \simeq M \circ_{\mathcal{O}'} N$ (Proposition 3.2.12), the map 6.3.3(i) is merely homotopic to a "fattened-up" version of the quotient map

$$M \circ N \to M \circ_{\mathcal{O}'} N$$

by coequalizing the left and right $\mathcal{O}'$ actions on N and M, respectively. Furthermore, item 6.3.3(ii) states that the functor ∂_* which sends a suitable $F \colon \mathsf{Alg}_{\mathcal{O}} \to \mathsf{Alg}_{\mathcal{O}'}$ to its sequence of derivatives $\partial_* F$ in fact takes values in $(\mathcal{O}', \mathcal{O})$ bimodules. From this perspective, there is an "evident" way to construct a functor $\mathsf{Alg}_{\mathcal{O}} \to \mathsf{Alg}_{\mathcal{O}'}$ from a sequence of derivatives: specifically, by the assignment $M \mapsto M \circ_{\mathcal{O}} (-)$. Let us denote by biRep the category of such functors F_M. Note that there is a natural inclusion $\mathsf{biRep} \hookrightarrow \operatorname{Fun}(\mathsf{Alg}_{\mathcal{O}}, \mathsf{Alg}_{\mathcal{O}'})$[14].

---

[13]That is, assuming that an operad structure can be described on $\partial_* \mathtt{Id}_{\mathsf{C}}$ and $\partial_* \mathtt{Id}_{\mathsf{D}}$ to begin with.

[14]We define here $\operatorname{Fun}(\mathsf{C}, \mathsf{D})$ latter is the category of finitary reduced simplicial functors $\mathsf{C} \to \mathsf{D}$.

We anticipate that a chain rule as in Conjecture 6.3.3 could further be used to prove an analogous result of Arone-Ching [3] in which they utilize their chain rule from [2] to show that the functor $F \mapsto \partial_* F$ which assigns a homotopy functor its sequence of Goodwillie derivatives admits a right adjoint Φ

$$\mathsf{Fun}(\mathsf{C}, \mathsf{D}) \underset{\Phi}{\overset{\partial_*}{\rightleftarrows}} \mathsf{BiMod}_{(\partial_* \mathrm{Id}_{\mathsf{D}}, \partial_* \mathrm{Id}_{\mathsf{C}})} \tag{6.3.4}$$

Here C and D are either Top_* or Spt and the explicit structure of Φ depends on choice of C, D. Arone-Ching further show that the that the adjunction (6.3.4) is *comonadic*[15], which essentially tells us that the Taylor tower of such a functor F may be recovered from its sequence of derivatives $\partial_* F$ as an appropriately sided bi-module, together with the action of the comonad $\partial_* \Phi$ via an equivalence

$$\mathrm{holim}_n P_n F \simeq \mathsf{Tot}\,\mathrm{Cobar}(\Phi, \partial_* \Phi, \partial_* F).$$

Crucial to their constructions is a precise model for $\partial_* F$ built from the derivatives of representable functors $\mathrm{Map}(X, -)$—we conjecture that a similar result holds for functors of structured ring spectra, where we use functors $F_M \in \mathsf{biRep}$ as our "building blocks".

6.3.5 An operad structure on the derivatives of the identity in a general model category

We additionally expect that the constructions in this document may be used as a general approach to describe an operad structure on the derivatives of the identity in a suitable model category (or perhaps ∞-category) C. What "suitable" means in this context is still a subject of ongoing matter, but we in particular would require (i) C be pointed and simplicial, (ii) the stabilization $\mathsf{Spt}(\mathsf{C})$ of C be closed, symmetric monoidal. We write $(\Sigma_{\mathsf{C}}^{\infty}, \Omega_{\mathsf{C}}^{\infty})$ for the

---

[15]See, e.g., [48] for a more detailed treatment of (co)monadic adjunctions.

stabilization adjunction between C and $\mathsf{Spt}(\mathsf{C})$.

Similarly to the cases described before, we begin with the cosimplicial resolution of Id_C via the stabilization adjunction. Either by analogous estimates from [15], [14] or via a modification of [2, 16.1], it should follow that $\partial_n \mathsf{Id}_\mathsf{C}$ can be constructed as the totalization of the cosimplicial diagram built from $\partial_n((\Omega_\mathsf{C}^\infty \Sigma_\mathsf{C}^\infty)^{k+1})$ for $k \geq 0$. Let us write K for $\Omega_\mathsf{C}^\infty \Sigma_\mathsf{C}^\infty$, and note that K admits a natural comonoid map $\mathsf{K} \to \mathsf{KK}$ via the unit $\mathsf{Id}_\mathsf{C} \to \Omega_\mathsf{C}^\infty \Sigma_\mathsf{C}^\infty$. A chain rule in the style of Conjecture 6.3.3) provides that the derivatives $\partial_* K$ of K be (at least up to homotopy) a cooperad[16] via the composite

$$\partial_* \mathsf{K} \to \partial_*(\mathsf{KK}) \simeq \partial_* \mathsf{K} \circ_{\partial_* \mathsf{Id}_{\mathsf{Spt}(\mathsf{C})}} \partial_* \mathsf{K}$$

In nice cases, this cosimplicial diagram is (weakly equivalent) to the cobar complex $C(\partial_* \mathsf{K})$ on $\partial_* \mathsf{K}$. However, it is more likely that this cooperad structure is defined only up to homotopy and is thus not rigid enough to provide a strict comparison of cosimplicial objects between $\partial_*((\Omega_\mathsf{C}^\infty \Sigma_\mathsf{C}^\infty)^{\bullet+1})$ and $C(\partial_* \mathsf{K})$.

One main benefit to our work is that we only need that the resultant diagram cosimplicial diagram admits a monoidal pairing with respect to $\mathring{\square}$. This would have to built by hand, but should be possible if we can provide a Snaith splitting of the counit K[17] or somehow otherwise find a way to produce a Tot-model for $\partial_* \mathsf{Id}_\mathsf{C}$ whose underlying cosimplicial object admits a $\mathring{\square}$-monoid structure. This is essentially our approach for $\mathsf{C} = \mathsf{Alg}_\mathcal{O}$, wherein we use a specific model for iterates of stabilization $UQ = \Omega_{\mathsf{Alg}_\mathcal{O}}^\infty \Sigma_{\mathsf{Alg}_\mathcal{O}}^\infty$ provided by [55] to obtain the diagram $C(\mathcal{O})$.

---

[16]Though this this does require some knowledge about the operad $\partial_* \mathsf{Id}_{\mathsf{Spt}(\mathsf{C})}$. For instance, when $\mathsf{C} = \mathsf{Top}_*$ this is just the identity symmetric sequence; for $\mathsf{C} = \mathsf{Alg}_\mathcal{O}$ this is the symmetric sequence with the ring spectrum $\mathcal{O}[1]$ concentrated at level 1.

[17]Compare with the Snaith splitting for $\mathsf{Alg}_\mathcal{O}$ is described in Section 3.3.1.

6.3.6 "Classifying operads" and homotopy descent

Once this operad structure has been established, it is reasonable to wonder how much of the original homotopy theory of C is captured by the category of algebras over $\partial_* \mathrm{Id}_C$ (a problem suggested by Haynes Miller). In the best case, we would obtain that C is equivalent to the category of algebras over $\partial_* \mathrm{Id}_C$. As suggested by the results of [15], [28], it is more likely this equivalence will hold only after restricting to a "nice" subclass of objects. We expect our box product pairings to provide a useful tool for obtaining such a result and outline one possible approach as follows.

The assignment $X \mapsto \Sigma_C^\infty X$, produces a functor from C to K-coalgebras whose right adjoint is given by taking "derived primitives". This adjunction is an equivalence is whenever the stabilization adjunction is comonadic (see [48]), which we conjecture holds precisely when restricted to elements of C such that the Taylor tower of Id_C converges. A Snaith splitting is key in translating the resultant K-coalgebra structure to a "Tate coalgebra" structure over a the cooperad $\partial_* \mathrm{K}$ [49]. This is part may be tricky, as even in the best known cases so far, the Snaith splitting only produces such an equivalence up to homotopy. However, this is precisely one benefit to our technique: We only need to show that the resulting diagram $C(\Sigma_C^\infty X)$ admits a (left) module structure over the underlying cosimplicial object of $\partial_* \mathrm{Id}_C$ with respect to the box product.

We summarize the above discussion in the following conjecture (see [26], [60, §6] for a similar discussion in ∞-categories)

Conjecture 6.3.7. *Let C be a "suitably nice" pointed simplicial model category, such that* Spt(C) *is symmetric monoidal. Then*

    (i) There is a natural model for the Goodwillie derivatives of the identity $\partial_* \mathrm{Id}_C$ as an operad and $\partial_* (\Sigma_C^\infty \Omega_C^\infty)$ as a cooperad which are Koszul dual.

(ii) There is a comparison map Ξ_C from C to $\partial_* \mathrm{Id}_C$-algebras in $\mathsf{Spt}(C)$ given by

$$\Xi_C \colon X \mapsto \mathrm{Tot}\, C(\Sigma_C^\infty X)$$

In particularly nice cases, we are interested in understanding when Ξ_C induces an equivalence of homotopy categories. One method for showing such an equivalence would be to provide a natural coaugmentation $\Phi_C(X) \to C(\Sigma_C^\infty X)$ (which may be possible only after localization, in which case Φ_C is essentially the localization functor), whose connectivity may be analyzed via the cubical analysis techniques developed in [33], [14], [15], [27]. In $\mathsf{Alg}_\mathcal{O}$, this coaugmentation is the unit map $X \to UQ(X)$ which induces an equivalence

$$X \simeq \mathrm{Prim}_{B(\mathcal{O})} \Sigma_{\mathsf{Alg}_\mathcal{O}}^\infty X \simeq X_{\mathsf{TQ}}^\wedge$$

for connective X [29]. In Top_* this coaugmentation is obtained (essentially) via the *Bousfield-Kuhn functor* [54], [20] (which factors through a suitable chromatic localization of spaces and lands in $T(n)$-local spectra), as shown by Heuts in [50].